BLACK AND OTHER
MINORITY PARTICIPATION
IN THE ALL-VOLUNTEER NAVY
AND MARINE CORPS

Major Industrial Research Unit Studies

The Wharton School's Industrial Research Unit has been noted for its "relevant research" since its founding in 1921. The IRU is now the largest academic publisher of manpower and collective bargaining studies. Publications include the Major Industrial Research Unit Studies, listed below, and monographs in special series, such as the Labor Relations and Public Policy Series, Manpower and Human Resources Studies, and Multinational Industrial Relations Series.

No. 49 Herbert R. Northrup et al., *Negro Employment in Southern Industry: A Study of the Racial Policies in the Paper, Lumber, Tobacco, Coal Mining, and Textile Industries*, Studies of Negro Employment, Vol. IV (1971) $13.50.

No. 50 Herbert R. Northrup et al., *Negro Employment in Land and Air Transport: A Study of Racial Policies in the Railroad, Airline, Trucking, and Urban Transit Industries*, Studies of Negro Employment, Vol. V (1971) $13.50.

No. 51 Gordon F. Bloom, Charles R. Perry, and F. Marion Fletcher, *Negro Employment in Retail Trade: A Study of Racial Policies in the Department Store, Drugstore, and Supermarket Industries*, Studies of Negro Employment, Vol. VI (1972) $12.00.

No. 53 Charles R. Perry, Bernard E. Anderson, Richard L. Rowan, and Herbert R. Northrup, *The Impact of Government Manpower Programs*, Manpower and Human Resources Studies, No. 4 (1975) $18.50.

No. 54 Herbert R. Northrup and Howard G. Foster, *Open Shop Construction* (1975) $15.00.

No. 55 Stephen A. Schneider, *The Availability of Minorities and Women for Professional and Managerial Positions, 1970-1985*, Manpower and Human Resources Studies, No. 7 (1977) $25.00.

No. 56 Herbert R. Northrup, Ronald M. Cowin, Lawrence G. Vanden Plas, et al., *The Objective Selection of Supervisors*, Manpower and Human Resources Studies, No. 8 (1978) $25.00.

No. 57 Herbert R. Northrup et al., *Black and Other Minority Participation in the All-Volunteer Navy and Marine Corps*, Studies of Negro Employment, Vol. VIII (1979) $18.50.

Order from the Industrial Research Unit
The Wharton School, University of Pennsylvania
Philadelphia, Pennsylvania 19104

BLACK AND OTHER MINORITY PARTICIPATION IN THE ALL-VOLUNTEER NAVY AND MARINE CORPS

(VOLUME VIII—STUDIES OF NEGRO EMPLOYMENT)

by

HERBERT R. NORTHRUP   STEVEN M. DIANTONIO
JOHN A. BRINKER   DALE F. DANIEL

and

RONALD M. COWIN   FRANK A. JENKINS III
EUGENE G. MATTISON III

INDUSTRIAL RESEARCH UNIT
The Wharton School
University of Pennsylvania

109494

This study was supported pursuant to the Organizational Research Programs, Office of Naval Research (Code 452) under contracts No. N0014-67-A-0216-0027 and No. N0014-76-G-0038. Since grantees preparing research with the assistance of government contracts are encouraged to express their own judgments freely, this study does not necessarily represent any official opinion or policy of the supporting agency. Moreover, the authors are solely responsible for the factual accuracy of all material developed in the study.

Reproduction in whole or in part permitted for any purposes of the
United States Government
All other rights reserved

Copyright © 1979 by the Trustees of the University of Pennsylvania
Library of Congress Catalog Number 78-72037
MANUFACTURED IN THE UNITED STATES OF AMERICA
ISBN: 0-89546-010-6

Foreword

This study is the eighth volume in the Studies of Negro Employment. The first six were published pursuant to grants from the Ford Foundation; the seventh was underwritten in part by the Ford Foundation, in part by the Maritime Commission, U.S. Department of Commerce, and in part by unrestricted grants to the Industrial Research Unit. Volume VIII, the first to deal with the government sector, was sponsored in part by the Office of Naval Research and in part by unrestricted funds provided by the industry members of the Industrial Research Unit's Research Advisory Group and the Labor Relations Council of the Wharton School. A complete list of these studies is found at the end of the book.

The decision to rely on all-volunteer armed services has meant that the procurement of manpower by the services is now dependent upon their ability to compete with private industry and other governmental services. It also means that their employment policies are being scrutinized to determine the extent to which they conform to national labor policies within the constraints of their institutional requirements.

This study is concerned with the racial policies of the Navy and Marine Corps. It seeks to determine the extent to which minorities have been integrated into the Navy and Marine Corps. It critically, but sympathetically, examines the manner in which these branches of the services recruit, train, upgrade, and retain minorities. The study is based upon extensive field work at Navy and Marine Corps installations over a four-year period in all sectors of the country and at all levels, plus a detailed examination of the literature and of the statistical record. The emphasis is on a comparison of Navy and Marine Corps practices and policies with those of private industry, again with due regard to the institutional constraints under which the armed services must perform their mission. The conclusion is that the

Navy and Marine Corps have come far toward equalitarian practices, but that both could profit by a greater emulation of private industry policies and programs.

The young authors who worked with me on this study are all veterans. Two, Dr. Steven M. DiAntonio and Mr. Dale F. Daniel, are graduates of the U.S. Naval Academy. Dr. DiAntonio received his Master of Business Administration and Doctor of Philosophy degrees from the Wharton School while studying under an Admiral Burke scholarship. He was forced to leave the Navy because of an injury and is now employed by a major consulting firm. Mr. Daniel received his Master of Business Administration degree at Wharton following his Navy service and is now working for a major corporation. Mr. John A. Brinker graduated from Juniata College and served six years as a naval cryptologic officer. He will receive his M.B.A. degree from Wharton in May 1979.

Mr. Ronald M. Cowin conceived the original study while working as a senior researcher for the Industrial Research Unit following his completion of the M.B.A. program. A former Marine Corps captain, he is now a personnel manager for a major corporation. Mr. Frank A. Jenkins III, now a certified public accountant, is also a Wharton M.B.A. recipient. After graduation from Morris Brown College, he served in the Air Force. Mr. Eugene G. Mattison III, the sixth Wharton graduate author of this study, served in naval intelligence following graduation from Brown University before he entered our graduate business school, and is now employed in the banking industry. Others who helped with fact gathering or data analysis are Messrs. Adolfo Castilla, Dale A. Miller, George Boomer, and Henry J. Pierce.

Numerous persons in the Navy, Marine Corps, the Department of Defense and in private industry were most helpful. We cannot note them all, but would like to mention a few who were especially helpful. First of all, the encouraging support of Dr. Glenn L. Bryan, Director, Psychological Services Division, Office of Naval Research, and his associates kept the project alive and contributed innumerable valuable suggestions. Over the years, excellent support and help was received from Lt. Tom

Foreword vii

McLaughlin, USN; Captain W. J. Loggan, USN; Colonel H. L. Blanton, Jr., and Lt. Colonel V. M. Salozzo, U.S. Marine Corps; Dr. David Robertson; and Mr. Edward Scarborough. Mr. Robert Krone, Vice-President, McDonnell-Douglas, and his staff were extremely helpful in providing relevant industry comparisons. Professors Howard E. Mitchell, Donald F. Morrison, and Richard L. Rowan at the Wharton School read the manuscript and made many valuable suggestions. The final manuscript was typed by Mrs. Jean McGrath, Mrs. Judith A. Pepper, and Mrs. Nancy E. Chiang, and edited by the Industrial Research Unit's Chief Editor, Mr. Robert E. Bolick, Jr., who also prepared the index. Mrs. Margaret E. Doyle, the Unit's Office Manager, handled the various administrative and accounting matters for the project.

> HERBERT R. NORTHRUP, *Director*
> Industrial Research Unit
> The Wharton School
> University of Pennsylvania

Philadelphia
November 1978

TABLE OF CONTENTS

	PAGE
FOREWORD	v

CHAPTER

I. INTRODUCTION	1
Definition of Terms and Goals	6

II. EVOLUTION TOWARD EQUAL OPPORTUNITY AND AFFIRMATIVE ACTION IN THE NAVY AND MARINE CORPS	8
History of Minority Personnel Involvement Policies: Revolutionary War Through Korean War	8
Civil War to Beginning of World War II	9
World War II to the Present	11
Executive Order 11246: Equal Opportunity vs. Affirmative Action	13
Affirmative Action: Confusion and Conflict	16
Recent Indicators of Racial Discontent	17
U.S.S. *Kitty Hawk,* October 1972	18
U.S.S. *Constellation,* November 1972	19
Impact of Incidents	21
Navy Race Relations Program: Phase I	22
Phase I Design	23
Command Affirmative Action Plans: Concept and Problems	25
Evaluation of Phase I	26
Navy Race Relations Program: Phase II	27
Converting RRES to EOPS	28
Phase II Design	28
Marine Corps Human Relations Program	30
Program Design	30

CHAPTER	PAGE
Navy and Marine Corps Servicewide Programs Contrasted	32
The Navy Affirmative Action Plan of 1976	33
Concluding Remarks	34
III. RECRUITMENT	36
Recruiting Organization	36
Line and Staff Relationships	37
The Recruiter: Selection, Training, and Responsibilities	40
Selection	40
Training	41
Job Responsibilities	43
Length of Assignment to Recruiting Duty	44
Recruiting Facilities and Facility Manning	45
Location	45
Minority Community Penetration Programs	46
Facility Manning for Recruiting Stations	47
Mental Testing: The Key to Enlisted Recruiting	48
Recruiting/Training Relationship	48
Mental Group Determination	49
Moral and Physical Criteria	49
Recruiting Mix	51
Service Entrance Standards and Minority Accessions	53
The Impact of the Educational Criterion	56
Educational and Mental Standards, Minority Accessions, and the Economy	57
Quality Impact	58
Project 100,000: A Case of Lower Mental Standards	60
Background	61
Evaluations of Project 100,000	61
Officer Recruiting	63
Educational Attainment	63
Entry Routes	65

Table of Contents

CHAPTER	PAGE

IV. A STATISTICAL ANALYSIS OF MINORITIES' UPGRADING OPPORTUNITIES IN THE ENLISTED NAVY 72

    Preservice Model 76
        Interpretation of Preservice Model Statistical Results 77
    Preservice Plus In-Service Model 81
        Interpretation of the Preservice Plus In-Service Model 82
    Separate Black and Ethnic Models 86
        Comparison of Original, Black, and Other Minority Personnel Models 86
    Cross-Sectional Model 88
        Cross-Sectional Model Interpretation 89
    Performance Evaluation Model 91
        Performance Model Interpretation 92
    Summary 94

V. OCCUPATIONAL CLASSIFICATION AND ASSIGNMENT 95

    Enlisted Minority Personnel Occupational Distribution 96
    The Classification and Assignment Process 99
        Volunteer Assignment 100
        Enlistment Options vs. Open Contract Assignment 102
        In-Fleet Formal Training Assignment 104
    Undesignated Strikers and On-the-Job Training 105
        Undesignated Striker Job Selection Process 106
        Problems with Job Selection and Training 107
        Recommendations for Upgrading Striker Placement Opportunities 108
    Industry Practices 109
    Formal Training Selection and Assignment 110

CHAPTER	PAGE
Selection for Formal School Training	110
Aptitude Testing: From AFQT and BTB to ASVAB	112
Minority Personnel Qualifications: The Problem of Academic Credentials	114
Minority Personnel Upgrading vs. Aptitude Testing	115
Mental Aptitude Inflation	116
Aptitude Test Validation	117
Effects of Sending Personnel Classified as Not School Eligible to Formal School	118
Alternative Selection and Training Methodologies	119
Self-Paced Training Methodology	122
Remedial Education	123
Current Remedial Programs	124
The JOBS Program	125
Is It Cost-Effective?	126
Officer Classification	127
Minority Officer Occupational Distribution	127
VI. ADVANCEMENT	134
Enlisted Minority vs. Nonminority Personnel Paygrade Distribution	134
Promotion Practices and Standards	136
Enlisted Advancement System Description	137
The Role of Performance Evaluations	140
The Role of Advancement Exams	141
Advancement Exam Domination of Promotion Selection	142
Alternative Testing Methods	145
Similar-Item-Difficulty Tests	145
Job-Sample Tests	146

Table of Contents xiii

CHAPTER PAGE

    Alternative Weighting of Promotion System Factors .. 147

    Military Justice 148

        Race-Ethnic Groups and Military Justice Offenses .. 149
        Causes of Disparate Offense Rates 150
        Race-Ethnic Groups, NJP Rates, and Organizational Effectiveness 150

    Lateral Transfer 151

    Manpower Control vs. Affirmative Action Promotion Programs 154

    Educational Opportunities and Career Counseling......... 155

        Educational Opportunities 155
        The BOOST Program 157
        Career Counseling 158

    Officer Promotions 159

        Promotion Standards and Practices 161
        The Importance of Career Pattern 162
        The Role of Fitness Reports 163

        Warfare Specialty and Military Occupational Specialty 165

VII. RETENTION 166

    Definition of Career 166

    Reenlistment Eligibility 167

        Reenlistment Eligibility vs. Occupational Specialty Manning Levels 169

    Minority vs. Nonminority Retention Situation: Statistical Analysis 171

        Interpretation of Model Results with respect to Minority Personnel Retention 171
        Minority Personnel Representation vs. Reenlistment Eligibility 173

    Reenlistment Incentives 174

CHAPTER	PAGE
Selective Conversion and Reenlistment (SCORE) Program	175
Selective Training and Reenlistment (STAR) Program	177
Guaranteed Assignment Retention Detailing (GUARD) Program	179
Officer Retention	180
Officer Incentives	181
Minority Officer Retention	181
Minority Officer Career Eligibility	183 |//
VIII. CONCLUDING REMARKS	185
Recruiting Organization	186
Recruitment and Classification Eligibility: Mental Aptitude Testing	187
Remedial Education	188
Alternative Training Methods: Self-Paced	192
Undesignated Strikers and On-the-Job Training	193
Promotion Eligibility	193
Advancement Exam Weighting	194
Alternative Policy for Promotion Selection	195
Alternative Testing	196
Performance Evaluations	197
Military Justice	198
Career Counseling	199
Retention	200
Reenlistment Eligibility	200
Lateral Transfer and Reenlistment-Induced Formal Training	201
Officers	202

Table of Contents

CHAPTER	PAGE
Minority Officer Recruiting	203
Officer Classification and Assignment	204
Officer Promotions	204
Officer Retention	206
Final Comment	206

APPENDIX

A. The Multiple Stepwise Regression Models: Statistical Technique ... 207

B. Preservice Model: Statistical Results ... 211

C. Preservice Plus In-Service Model: Statistical Results ... 217

D. Separate Black and Other Minority Personnel Models: Statistical Results ... 221

E. Cross-Sectional Model: Statistical Results ... 225

F. Performance Evaluation Model: Statistical Results ... 229

INDEX ... 235

LIST OF TABLES

TABLE **PAGE**

CHAPTER I

1. Navy and Marine Corps Officer and Enlisted Personnel, Participation Rates by Race 3
2. Percentage Distribution of Officer Personnel by Race and Rank 4
3. Percentage Distribution of Enlisted Personnel by Race and Paygrade 5

CHAPTER III

1. Mental Group Classification and AFQT Score 50
2. Navy Enlisted Accessions by Race and Mental Group .. 54
3. Marine Corps Enlisted Accessions by Race and Mental Group 55
4. Percentage Distribution of Navy and Marine Corps Accessions by Race and Mental Group 57
5. Navy and Marine Corps Black Enlisted Accessions as a Percentage of Total Accessions: Unemployment Rate as a Percentage of Total Civilian Work Force 59
6. Employed Persons in Selected Professional Occupations by Race 64
7. Minority Officer Accessions by Source of Procurement (Navy) 68
8. Minority Officer Accessions by Source of Procurement (USMC) 69

CHAPTER IV

1. Definition of Advancement Function Variables 73

xvi

List of Tables xvii

TABLE		PAGE

2. Preservice Model Statistics: Significant Variables Listed in Order of Significance 78

3. Preservice Plus In-Service Model Statistics: Significant Variables Listed in Order of Significance .. 83

CHAPTER V

1. Percentage Distribution of Enlisted Men by Race and Occupational Group in the Navy and Marine Corps 97

2. Total Officer Personnel by Race and Occupational Group in the Navy and Marine Corps 128

3. Selected Professional Occupations by Race in the Navy and Marine Corps 130

4. U.S. Naval Academy Program Accessions 133

CHAPTER VI

1. Total Enlisted Personnel by Race and Paygrade in the Navy and Marine Corps 135

2. Weighting of Navy Enlisted Advancement Factors for Final Multiple Score Computation 138

3. Percentage Distribution of Officer Personnel by Race and Paygrade in the Navy and Marine Corps 160

4. Time-in-Grade Requirements for Promotion and Approximate Percentage of Officers in Selection Zone 163

CHAPTER VII

1. Enlisted Retention and Reenlistment Race-Ethnic Comparisons 174

2. Navy Officer Retention by Warfare Community 182

LIST OF FIGURES

FIGURE   **PAGE**

CHAPTER II

1. Navy Equal Opportunity Goals and Objectives ... 15
2. Marine Corps Equal Opportunity Goals ... 16

CHAPTER III

1. Navy Personnel Management and Recruiting Command Organization ... 37
2. Marine Corps Recruiting Command Organization ... 38
3. Navy Recruiting Criteria ... 51
4. Overview of Navy and Marine Corps Officer Entry Routes ... 66

CHAPTER V

1. Prevolunteer Service Hierarchy of Assignment Objectives ... 101
2. The Basic Test Battery (BTB) ... 111

CHAPTER VI

1. Composite Score Computation for Marine Corps Promotion to Corporal (E-4) and Sergeant (E-5) ... 139
2. Typical Career Pattern for Surface Warfare Officer ... 164

APPENDIX B

1. Preservice Plus In-Service Correlation Matrix ... 213

APPENDIX D

1. Black Model Statistics: Significant Variables Listed in Order of Significance ... 224

List of Figures

FIGURE PAGE

2. Other Minority Model Statistics: Significant Variables Listed in Order of Significance 224

APPENDIX E

1. Cross-Sectional Model Statistics: Significant Variables Listed in Order of Significance 227

APPENDIX F

1. Performance Evaluation Model Statistics: Significant Variables Listed in Order of Significance 231

2. Performance Evaluation Model Correlation Matrix 232

CHAPTER I

Introduction

Throughout the histories of the Navy and the Marine Corps, minorities have been underrepresented in both services. Today, the all-volunteer armed forces is a reality, making it necessary for the armed services to compete with private industry for manpower. In this competition, the Department of Defense and the individual services realize, as private industry realizes, that due regard must be given to the public policy of opening doors previously closed to minorities. Of the laws affecting equal opportunity policies today, the most important is Title VII of the Civil Rights Act of 1964, as amended in 1972. The federal government, however, in dealing with employment discrimination, has an additional tool, Executive Order 11246, which requires employers operating under its jurisdiction to avoid discrimination and to take affirmative action to ensure the absence of discrimination in all phases of employment. In 1970, this order was extended to cover the Navy and Marine Corps, as well as the other armed services. To be in concert with these laws as they have been interpreted, the services' public policies must afford minorities every available opportunity to develop their abilities and to achieve the highest position commensurate with those abilities, and must aim at making minorities' representation in all positions reflect those minorities' representation in the nation.

Consequently, the avowed policy of the United States Navy and Marine Corps is "to insure equality of opportunity and treatment for all military members and civilian employees of the Department of the Navy, regardless of race, creed, color, sex or national origin." [1] Further, in response to Department of Defense Directive 1100.15 ("Equal Opportunity Within the Department of Defense"),[2] both services are attempting

---

[1] Chief of Naval Operations, *U.S. Navy Equal Opportunity Manual*, OPNAV Instruction 5354.1A (Washington, D.C.: Department of the Navy, 1978), p. 2; and comparable Commandant of the Marine Corps, "Equal Opportunity Within the Marine Corps," Marine Corps Order 5350.5A (Washington, D.C.: U.S. Marine Corps, Headquarters, 1974), p. 2.

[2] This directive was originally issued on December 14, 1970, and since reissued on June 3, 1976, under the title "Department of Defense Equal Opportunity Program."

> To increase and intensify ... efforts to attain and retain the highest quality officer and enlisted volunteers from all segments of society, seeking to achieve increased representation of minority personnel in the various categories and grades of the service which is proportional to the demography of the source populations. ...
>
> To identify and eliminate all bias, i.e., insure equal opportunity for: selection for programs, appointments or promotion; classification to occupational fields; technical and professional schooling; developmental experiences and progression in duty assignment; performance evaluations; pro-pay; advancement and promotion; retention, reenlistment and career status; etc. ...
>
> To achieve and guarantee legal and administrative processes which are responsive to minority as well as majority personnel. ...
>
> To conduct workshops, conferences and educational, recreational, and social programs to enhance interracial understanding, cooperation and respect among all ... personnel.[3]

During the past thirty years, the two services have achieved some success. In 1950, 10.0 percent of the United States' population was black;[4] but in 1949, only 4.4 percent of Navy personnel and 2.5 percent of Marine Corps personnel were black. Today, the percentage of blacks in the country's population is approximately 11 percent.[5] As of June 30, 1977, 8.5 percent of enlisted Navy personnel and 17.3 percent of enlisted Marine Corps personnel were black. In addition, from 1949 to 1976, black representation in the Navy's and Marine Corps' officer personnel increased considerably (see Table I-1).

Despite this improvement, both services still have a problem with minority participation. The issue of minority participation and equal opportunity goes deeper than the mere size of minority representation in the services. The proper degree of participation and the dispersion of that participation are prerequisites for complete integration of the services. The distribution, or frequency, of minorities throughout the hierarchy of occupational skill levels and paygrades reflects directly the degree to which

---

[3] Chief of Naval Operations, *U.S. Navy Equal Opportunity Manual*, OPNAV Instruction 5354.1 (Washington, D.C.: Department of the Navy, 1974), pp. 2-4.

[4] U.S., Department of Commerce, Bureau of the Census, *United States Census of Population: 1950*, Vol. II, *Characteristics of the Population*, Pt. 1, United States Summary.

[5] U.S., Department of Commerce, Bureau of the Census, *United States Census of Population: 1970*, *General Population Characteristics*, PC(1)-B1, United States Summary.

Introduction

TABLE I-1

Navy and Marine Corps Officer and Enlisted Personnel Participation Rates by Race 1949, 1964-1977

	Navy Enlisted Personnel Percentage Negro	Navy Officer Personnel Percentage Negro	Marine Corps Enlisted Personnel Percentage Negro	Marine Corps Officer Personnel Percentage Negro
Year				
1949	4.40	0.01	2.50	0.01
1964	6.00	0.30	8.70	0.40
1965	5.80	0.30	8.90	0.40
1966	5.10	0.30	9.40	0.70
1967	4.70	0.30	10.30	0.70
1968	5.00	0.40	11.50	0.90
1969	5.40	0.70	11.60	1.20
1970	5.40	0.70	11.20	1.30
1971	5.40	0.70	11.40	1.30
1972	6.40	0.90	13.70	1.50
1973	7.70	1.10	16.90	1.90
1974	8.40	1.30	18.10	2.40
1975	8.00	1.40	18.10	3.00
1976	8.00	1.60	17.00	3.40
1977	8.50	1.80	17.30	3.60

Sources: 1949 to 1970 figures—Office of the Deputy Assistant Secretary of Defense (Equal Opportunity), *The Negro in the Armed Forces: A Statistical Fact Book* (Washington, D.C.: Department of Defense, 1971), pp. 1, 4-7. 1971 to 1977 figures—Defense Manpower Data Center, Military Master Files as of July 30, 1978.

Note: The 1949 to 1970 figures are as of December 31 of each year. The 1971 to 1977 figures are as of June 30 of each year.

minorities have a share in the more highly skilled and technical job categories and in the more senior paygrades.

As Table I-1 clearly demonstrates, black participation in the enlisted ranks far exceeds that in the officer ranks. Table I-2 shows the percentage distribution of officer members by race and rank and reveals that blacks are underrepresented in the higher

TABLE I-2
Percentage Distribution of Officer Personnel
by Race and Rank as of
December 31, 1977

Officer Rank	Navy Total Officers	Navy Black	Marine Corps Total Officers	Marine Corps Black
0-7+	0.4	0.2	0.4	—
0-6	5.8	1.8	3.2	0.3
0-5	12.0	3.2	8.0	0.9
0-4	19.7	7.4	14.4	4.6
0-3	27.5	30.5	24.3	19.1
0-2	15.5	21.9	25.5	38.7
0-1	14.3	27.1	18.1	27.3
W1-W4	4.8	7.9	6.1	9.1
TOTAL	100.0	100.0	100.0	100.0

Sources: Bureau of Naval Personnel (Pers-61), "Navy Wide Demographic Data Base for First Quarter FY-78 (1 Oct 77 to 31 Dec 77)" (Washington, D.C.: Department of the Navy, 1978); U.S. Marine Corps, Headquarters, Manpower Planning, Programming and Budgeting Branch, August 1978.

ranks and are overrepresented in the lower ranks. For example, in the Marine Corps, 11.6 percent of all officers are lieutenant colonels (0-5) or higher, while only 1.2 percent of the Marine Corps' black officers have achieved this rank. In the Navy, 34.6 percent of the officer corps are in rank 0-2 or lower, while 56.9 percent of black officers fall into that category.

As might be expected, this disproportionate representation occurs in the nine enlisted paygrades for the Navy and Marine Corps. Table I-3 shows that, in the Navy, about 41 percent of all enlisted and only 28 percent of enlisted blacks are found in paygrades E-5 through E-9; in the Marine Corps, over 30 percent of all enlisted and only 26 percent of enlisted blacks are in these paygrades. In the non-petty officer grades (E-1 to E-3) are found 42 percent of the Navy's enlisted and more than 55 percent of its enlisted blacks; 53 percent of the Marine Corps' enlisted and almost 60 percent of its enlisted blacks fall into the same paygrades.

TABLE I-3
Percentage Distribution of Enlisted Personnel
by Race and Paygrade as of
December 31, 1977

	Navy		Marine Corps	
Enlisted Paygrade	Total Enlisted Personnel	Black Enlisted Personnel	Total Enlisted Personnel	Black Enlisted Personnel
E-9	0.80	0.40	0.70	0.40
E-8	1.80	1.10	1.90	1.50
E-7	6.60	4.40	5.00	3.80
E-6	14.00	8.30	8.20	6.70
E-5	17.61	13.40	14.60	13.50
E-4	17.60	17.10	17.10	14.50
E-3	20.90	25.70	22.80	23.30
E-2	11.40	16.00	16.80	19.50
E-1	9.30	13.60	12.90	16.80
TOTAL	100.00	100.00	100.00	100.00

Sources: Bureau of Naval Personnel (Pers-61), "Navy Wide Demographic Data Base for First Quarter FY-78 (1 Oct 77 to 31 Dec 77)" (Washington, D.C.: Department of the Navy, 1978); U.S. Marine Corps, Headquarters, Manpower Planning, Programming and Budgeting Branch, August 1978.

The lack of proportionate distribution of minorities in the enlisted and officer ranks and in the paygrades reflects a lack of proportionate distribution of minorities in occupational specialties. Minorities are underrepresented in the more skilled enlisted occupations and in all officer occupational specialties. Minority officers tend to be in nonprofessional occupational specialties and not in top leadership positions.

Thus, despite their statistical improvement over the last decade, the Navy and Marine Corps have much to accomplish before they can satisfy Executive Order 11246 and can claim to have effectively implemented their avowed policies on minority participation in their services. What they must accomplish and how it can be accomplished are the subject of this study.

DEFINITION OF TERMS AND GOALS

In this study, the terms *equal opportunity, affirmative action, upgrading,* and *mobility* have been used so far and will be used frequently. The first two terms are subject to popular misconceptions, and the last two have particular applications to manpower; therefore, before outlining the goals of this study, we must define these crucial terms.

A program of *equal opportunity* deals with eliminating any racial bias in an institution's criteria for hiring, job classification, and promotion selection. Such a program is the sort to which the *Navy Equal Opportunity Manual* and the Marine Corps' order "Equal Opportunity Within the Marine Corps" commit both services. The implementation of such a program alone, however, does not meet the commitments to affirmative action which are broadly outlined in DOD Directive 1100.15 and in the regulations of Executive Order 11246. A program of *affirmative action* deals with recruiting and upgrading minority members of an institution so that they are represented more proportionately in all paygrades and occupational categories. *Upgrading* is defined here as a process which provides educational and vocational training to disadvantaged individuals to improve their employment/promotion opportunities and to facilitate these individuals' advancement to higher paygrades.[6] *Mobility,* a broad concept, refers to the general movement of employees from one job to another.

This study is concerned with minority upgrading and mobility in the Navy and Marine Corps.[7] The study analyzes and evaluates the Navy's and Marine Corps' policies on recruitment, occupational classification, assignment, advancement, retention, and lateral transfer, as well as their human relations programs. Our analysis and evaluation involve, of course, a look not only at the

---

[6] For manpower study purposes, a *disadvantaged person* is defined as a poor person who does not have suitable employment and is either (1) a school dropout, (2) a member of a minority group, (3) under 22 years of age, (4) 45 years of age or older, or (5) handicapped. G. F. Bloom and H. R. Northrup, *Economics of Labor Relations,* 7th ed. (Homewood, Illinois: Richard D. Irwin, Inc., 1973), p. 476.

[7] Every effort was made to obtain the relevant data on blacks, Indians, Malayans, and Mongolians and, when necessary, to differentiate between these races. When all this material was not available, but data on blacks were, these data were used and conclusions drawn, when appropriate, concerning all minorities.

Introduction

impact on minorities of these institutional policies, procedures, and programs but also at their legal aspects and ramifications.

Finally, this study offers recommendations on better means for selecting disadvantaged minority personnel with the *capabilities* for successful upgrading. Recommendations are also made on how to upgrade competent, disadvantaged personnel (of which minorities constitute a disproportionate percentage) through the development of an operational human resource management philosophy.

As the largest vocational training institution in the United States, the military can contribute significantly toward the development of the nation's human resources. Implementing an effective human resource management philosophy can provide a more productive individual and, therefore, benefit both the individual and the military through his or her increased contribution to the organization. The importance of not compromising the primary mission of the services, however, must be kept in mind. Therefore, in making our recommendations, we take into account the services' need for quality in their personnel and for cost-effectiveness in their programs.

CHAPTER II

Evolution Toward Equal Opportunity and Affirmative Action in the Navy and Marine Corps

The Navy and the Marine Corps have altered their policies on minority participation throughout their histories in response to the mores of the times. This, of course, is to be expected in a democratic society in which the armed forces reflect societal changes in attitudes and opinions. To understand the institutional environment within the services today, we turn to a brief history of the evolving racial policies of the Navy and the Marine Corps.

HISTORY OF MINORITY PERSONNEL INVOLVEMENT POLICIES: REVOLUTIONARY WAR THROUGH KOREAN WAR

The navy of the Revolutionary War was in desperate need of men. Shipboard service was notorious for its brutality and its execrable living conditions. Mass desertions were not uncommon and created manpower shortages requiring ship commanders to rely on blacks as a last resort to help man the ships. Additionally, black coastal pilots were much in demand because of their knowledge of the coastal waters. Promises of emancipation, as well as the usual patriotic motives and promises of money and land, prompted blacks to serve. Because manpower data of the period do not generally include race or ethnic information, accurate estimates of the number of blacks serving in the Continental navy are not available.[1]

---

[1] Dennis D. Nelson, *The Integration of the Negro into the U.S. Navy* (New York: Farrar, Straus and Young, 1951), p. 1.

Official government policy barred blacks from the armed services after the Revolution.[2] Congress prohibited blacks from entering the state militias in 1792, and Benjamin Stoddert, the first secretary of the navy, barred blacks and mulattoes from the Navy and Marine Corps.[3] Despite these official policies, blacks served in the 1798-1800 naval war with France and in the War of 1812, presumably because of military manpower needs.[4] They constituted approximately one-sixth of the Navy's enlisted ranks during the War of 1812 and were integrated into all shipboard occupational ratings.[5] When the War of 1812 ended, so did the services' need for additional manpower. Official proscriptions against black enlistments were more closely followed than they were during the war. Few blacks served in the Mexican War, and from 1848 until the Civil War, the Navy was essentially an all-white service.[6]

Civil War to Beginning of World War II

The Civil War brought with it a new demand for increased military manpower. Although the services were experiencing difficulty in raising the needed manpower, many Northern whites were opposed to using blacks in the armed forces. Additionally, President Lincoln was intent on maintaining border state loyalty and would not sanction black enlistments in the army for fear of having the conflict perceived as an abolitionist war.[7] Despite these obstacles, the Navy again officially authorized black enlistments without service restrictions in September 1861.[8]

As the war dragged on, the need for military manpower increased. Finally, with the issuance of the Emancipation Proclama-

---

[2] Richard M. Dalfiume, *Desegregation of the U.S. Armed Forces: Fighting on Two Fronts, 1939-1953* (Columbia, Mo.: University of Missouri Press, 1969), p. 6.

[3] Richard J. Stillman, II, *Integration of the Negro in the U.S. Armed Forces*, Praeger Special Studies in U.S. Economic and Social Development (New York: Frederick A. Praeger, Publishers, 1968), p. 8.

[4] Dalfiume, *Desegregation of the U.S. Armed Forces*, p. 6.

[5] George Brooks Harrison, "An Analysis of Racial Integration in the Armed Forces of the United States of America" (Master's thesis, University of Pennsylvania, 1970), p. 32.

[6] Stillman, *Integration of the Negro*, p. 9.

[7] Dalfiume, *Desegregation of the U.S. Armed Forces*, p. 6.

[8] Nelson, *The Integration of the Negro*, p. 5.

tion on January 1, 1863, the Union abandoned attempts to assuage proslavery factions and began to encourage black enlistment in all the services.[9] Blacks were allowed to enlist in all general service ratings in the Navy, and many distinguished themselves in battle. Robert Smalls, a slave-pilot, pirated the Confederate gunboat *Planter* away from Charleston harbor to a Union port.[10] He later served as the *Planter*'s captain under the Union flag.[11]

The Southern states' governments were controlled by the politicians of the North from 1865 to 1876. During this period, some blacks enjoyed positions of power in the South as a result of the carpetbagger governments. Several black congressmen were elected, and black militiamen were appointed. As control of the South gradually fell back into the hands of Southerners, however, blacks lost much, if not all, of the status they had gained. Although no longer slaves, they continued to be economically and politically dependent upon white society. This situation was not peculiar to the South. Discrimination was the rule rather than the exception throughout the country. It took on many forms and existed in varying degrees. Segregation was institutionalized by the Supreme Court's *Plessy* v. *Ferguson* decision (1896), which established the "separate but equal" doctrine.[12]

The Navy continued its policy of allowing blacks to serve in all ratings during the post-Civil War era on an integrated basis. Blacks served only as enlisted men, however, and most often in the lowest ranks.[13] Black sailors continued to serve on an integrated basis during the Spanish-American War and distinguished themselves in battle at Manila and Santiago.[14]

The radical Populists gained power in the South during the early twentieth century. They were strongly antiblack and encouraged rigid social segregation. The executive and judicial branches of the federal government often acquiesced to the Populist demands, and segregation rapidly became an official, as well as informal, way of life in America.[15]

---

[9] Stillman, *Integration of the Negro*, p. 10.

[10] Nelson, *The Integration of the Negro*, p. 5.

[11] Harrison, "An Analysis of Racial Integration," p. 34.

[12] *Plessy* v. *Ferguson*, 163 U.S. 537 (1896).

[13] Harrison, "An Analysis of Racial Integration," p. 37.

[14] Nelson, *The Integration of the Negro*, p. 6.

[15] Stillman, *Integration of the Negro*, p. 12.

Official Navy policy at the beginning of World War I still allowed blacks to enlist in all service ratings, but informal Navy practices began to align themselves with the segregationist influences of the society at large. In 1919, for the first time in its history, the Navy instituted a segregationist policy. This policy restricted blacks to serve in the messman or steward branch.[16] Ten thousand black recruits served the Navy in this capacity in World War I,[17] while a few holdovers from earlier periods still served in engineering and general ratings.

Navy policy, after World War I, reflected the state of race relations in the country as a whole. Black personnel not only continued to be assigned only to the messman rating, but, in addition, the enlistment of black sailors was almost completely discontinued in the postwar force reduction as large numbers of Philippine nationals were recruited to fill the messman branch from 1919 to 1932. During the military buildup in the 1930s, blacks were again actively recruited, but again they were allowed to enlist only in the messman branch.[18]

World War II to the Present

With World War II came still another military manpower shortage. The Selective Service Law of September 1940 called citizens to service without regard to race, color, or creed. Because the Navy at this time still relied exclusively on volunteers, it was able to continue enlisting blacks only as messmen. In April 1942, the Navy relaxed its racial restrictions on service ratings, but blacks were still trained in segregated units and, except as messmen, were not assigned to seagoing vessels. Furthermore, no black officers were commissioned.

In 1940, the War Department announced that black accessions would be increased such that blacks would constitute roughly the same proportion of the armed forces as they did of the general population—approximately 10 percent. In March 1943, the War Manpower Commission began pressuring the Navy to meet this goal. As black recruits increased in number, it became apparent to the Navy that they could not all be assigned to shore stations. Various "experiments" were tried over the next two years, including the assignment of all-black crews to two antisubmarine

---

[16] Nelson, *The Integration of the Negro*, p. 8.

[17] Stillman, *Integration of the Negro*, p. 16.

[18] Nelson, *The Integration of the Negro*, pp. 10, 11.

vessels; and finally, on April 13, 1945, the Navy allowed blacks to serve on auxiliary ships in unrestricted capacities and later that year integrated its basic training facilities.[19]

It was not until after the war that all racial restrictions were removed in the Navy. In February 1946, the Navy lifted all service restrictions on blacks and ordered the complete integration of all Navy facilities. This move can be attributed to the strength and foresight of then Secretary of the Navy James Forrestal. At this time, the other services still had segregationist policies, and no direct congressional or executive desegregation orders had been issued to the services.

The Marine Corps opened its ranks to blacks for the first time in its history in May 1942. Segregated training facilities were provided at Montford Point, North Carolina. Basic training began for the first black marines in August 1942.[20] After basic training, the new black marines were organized into segregated defense battalions. The purpose of these battalions was to occupy the Pacific islands and atolls already captured from the Japanese in order to free more white marines for new assaults. As the war front was pushed further across the Pacific, defense battalions were no longer needed to secure the positions captured earlier in the war. Blacks were then organized into special ammunition and supply companies to help support combat troops during amphibious landings. Black participation reached its peak in September 1945, when 17,119 blacks were in Marine Corps uniform.[21]

After the war, the number of black marines dropped off sharply. By December 1946, there were only 2,238 enlisted blacks serving in the Corps and no black officers. In the spring of 1947, black marines were given their choice of either being discharged or transferred to the stewards branch.[22]

A year later, President Harry S. Truman issued Executive Order 9981 of July 26, 1948.[23] This order called for "equality of treatment and opportunity for all persons in the armed ser-

---

[19] *Ibid.*, pp. 19-20.

[20] "The Negro in the Marine Corps" (Washington, D.C.: U.S. Marine Corps, Headquarters, n.d.), p. 5.

[21] *Ibid.*, p. 12.

[22] *Ibid.*

[23] Executive Order No. 9981, *Federal Register* 13, No. 146, July 28, 1948, p. 4313.

vices without regard to race, color, religion or national origin." [24] This executive order did not push for reform to the same degree as Navy Secretary James Forrestal's desegregation order to the Navy issued two years earlier.

Although Executive Order 9981 did not explicitly require desegregation, it had two important consequences. First, it established the President's Committee on Equality of Treatment and Opportunity in the Armed Services. This committee, chaired by Charles H. Fahy, served as a formal symbol of the president's interest in improving the status of black servicemen. It also facilitated long overdue discussions within the services concerning alternatives to segregation.[25] Second, Louis B. Johnson, who succeeded Forrestal as secretary of defense in 1949, interpreted the executive order as requiring desegregation and ordered the services to submit integration plans.

The Korean War helped to speed the integration process. The demand for military manpower again increased. The segregated services found themselves with a shortage of whites on the front line and an excess of blacks in rear supply units. The need for replacement troops made the integration of combat units a sound military policy. Toward the end of the war, the Navy was receiving blacks at a rate of 4.3 percent of total enlistments, and the Marine Corps, at 8.0 percent.[26]

EXECUTIVE ORDER 11246: EQUAL OPPORTUNITY VS. AFFIRMATIVE ACTION

President Lyndon B. Johnson issued Executive Order 11246 in 1965, requiring affirmative action by government contractors. The secretary of labor then created the Office of Federal Contract Compliance (now the Office of Federal Contract Compliance Programs) to carry out this governmental function. The OFCC authored affirmative action program requirements for nonconstruction contractors through its General Order No. 4.

> An acceptable affirmative action program must include an analysis of areas within which the contractor is deficient in the utilization of minority groups and women, and further, goals and timetables to which the contractor's good faith efforts must be directed to correct the deficiencies and, thus to achieve prompt and full utiliza-

---

[24] *Ibid.*

[25] Stillman, *Integration of the Negro*, p. 44.

[26] *Ibid.*, Table 3, p. 66.

tion of minorities and women, at all levels in all segments of his work force where deficiencies exist.[27]

Executive Order 11246 and its implementing regulations were extended to cover the Navy and Marine Corps, as well as the other services, by the *Code of Federal Regulations* in 1970.[28] In compliance with the executive order, the Navy and Marine Corps have established affirmative action programs, as well as equal opportunity programs.

The Navy and Marine Corps are striving to attain overall proportionate minority representation and to distribute minorities proportionately across paygrades and occupations. Both the Navy[29] and Marine Corps[30] have produced equal opportunity manuals, which present the equal opportunity goals and objectives of each service, as well as policy and directive statements for the achievement of these goals. (See Figures II-1 and II-2 for a complete statement of the equal opportunity goals of the Navy and Marine Corps, respectively.) Below are two interrelated objectives pertaining to minority recruiting, upgrading, and mobility which the Navy and Marine Corps are striving to attain:

1. to intensify efforts to attain and retain the highest quality volunteers from all segments of society while seeking to achieve increased minority personnel representation which is proportional to the demography of the population;
2. to insure equal opportunity for: selection into programs, advancements and promotion, classification into occupational fields, technical and professional schooling, progression in duty assignment, performance evaluations, retention, reenlistment and career status, and living and working conditions.[31]

---

[27] U.S., Department of Labor, Equal Employment Opportunity, Office of Federal Contract Compliance, "Affirmative Action Plans," *Federal Register* 39, No. 32, February 14, 1974, p. 5630.

[28] 32 *Code of Federal Regulations* § 191.1 (1976).

[29] Chief of Naval Operations, *U.S. Navy Equal Opportunity Manual*, OPNAV Instruction 5354.1 (Washington, D.C.: Department of the Navy, 1974); superseded by OPNAV Instruction 5354.1A (1978).

[30] Commandant of the Marine Corps, "Equal Opportunity Within the Marine Corps," Marine Corps Order 5350.5A (Washington, D.C.: U.S. Marine Corps, Headquarters, 1974).

[31] Chief of Naval Operations, OPNAV Instruction 5354.1, pp. 2, 3; and Commandant of the Marine Corps, Marine Corps Order 5350.5A.

FIGURE II-1
Navy Equal Opportunity Goals and Objectives

Goal I. To attract to the Navy people with ability, dedication, and capacity for growth. Specifically, the Navy must be able to obtain the very best talent available in this nation regardless of race, religion, creed, sex, economic background, or national origin.

(1) Objectives

(a) To increase and intensify the Navy's efforts to attain and retain the highest quality officer and enlisted volunteers from all segments of society, seeking to achieve increased representation of minority personnel in the various categories and grades of the service which is proportional to the demography of the source populations.

(b) To create and maintain a Navy climate of equal opportunity and treatment for all people regardless of race, creed, religion, sex, or national origin.

(c) To establish educational programs within the Navy and in association with the Navy (e.g., NJROTC, Sea Cadets, Community Service Programs) to assist persons in attaining a common educational base which will aid them in achieving a level at which they can compete equitably with their peers.

(d) To establish open recreational programs within the Navy and in association with the Navy which will aid individuals to achieve, maintain, and support Navy mission-oriented goals of combat readiness.

(e) To establish open social programs which will create and maintain an atmosphere of personnel and mission integration within the Navy structure and in association with the Navy structure, i.e., NJROTC, Sea Cadets, Community Service Programs.

Goal II. To provide real opportunity for all personnel of the Department of the Navy to rise to as high a level of responsibility as possible, dependent only on individual talent and diligence.

(1) Objective

(a) To identify and eliminate all bias, i.e., insure equal opportunity for: selection for programs, appointments or promotion; classification to occupational fields; technical and professional schooling; developmental experiences and progression in duty assignment; performance evaluations; pro-pay; advancement and promotion; retention, reenlistment and career status; etc.

Goal III. To make service in the Department of the Navy a model of equal opportunity for all regardless of race, creed, religion, sex, or national origin, i.e., the Navy must strive to elevate the dignity of each individual and eliminate all vestiges of discrimination and intolerance so that all members of the naval service can be equally proud to serve.

(1) Objectives

(a) To create and insure equal opportunity in living and working conditions in the Navy community.

(b) To achieve and guarantee legal and administrative processes which are responsive to minority as well as majority personnel.

(c) To conduct workshops, conferences and educational, recreational, and social programs to enhance interracial understanding, cooperation, and respect among all Naval Personnel.

Source: Chief of Naval Operations, *U.S. Navy Equal Opportunity Manual*, OPNAV Instruction 5354.1 (Washington, D.C.: Department of the Navy, 1974).

FIGURE II-2
Marine Corps Equal Opportunity Goals

The following equal opportunity goals are established in order to expedite the attainment of true equal opportunity throughout the Marine Corps.

1. To achieve an equitable ethnic group balance in each DOD Occupational Group through qualified minority personnel accessions so that the percentage of minority personnel in each group approximates the percentage of minority personnel in the Marine Corps.
2. To achieve an equitable racial group balance in major Fleet Marine Force commands through personnel assignments so that the percentage of minority personnel in each command approximates the percentage of minority personnel in the Fleet Marine Force.
3. To increase the number of minority officer accessions so that by 1980 the percentage of these accessions will approximate the percentage of minority college enrollments.
4. To ensure fair, impartial and correct justice for all personnel and to dispel any doubts Marines may have about the Uniform Code of Military Justice.
5. To eliminate cultural bias from current testing documents and entry level assignments.
6. To reduce the rate of racial incidents by sensitive leadership, proper actions, and the development of an awareness on the part of all Marines.
7. To ensure public affairs efforts provide for appropriate coverage of the minority community, including general officer speaking engagements.
8. To achieve and maintain equal opportunity in all off-base housing.

Source: Commandant of the Marine Corps, "Equal Opportunity Within the Marine Corps," Marine Corps Order 5350.5A (Washington, D.C.: U.S. Marine Corps, Headquarters, 1974).

Affirmative Action: Confusion and Conflict

Until recently, the Navy and Marine Corps, in a real sense, did not practice affirmative action. They did not affirmatively seek, as a matter of regular practice, to aid minorities in overcoming labor market barriers, nor were their recruiting practices constituted to be able to do so.

Field interviews have indicated that few persons within the Navy have a good understanding of the concept of affirmative action.[32] Often, affirmative action and equal opportunity were used synonymously. In some instances, doing something in the minority community which was normally and regularly done in the majority community was considered affirmative action. For example, one person interviewed pointed out, as an example of

---

[32] Field interviews were conducted at recruiting commands in the following cities between 1973 and 1976: Philadelphia, Pa.; Arlington, Va.; San Diego, Calif.; Norfolk, Va.; Atlanta, Ga.; Washington, D.C.; Oklahoma City, Okla.; San Francisco, Calif.; and Pensacola, Fla.

affirmative action, the Navy's implementation of Junior Reserve Officer Training Corps (JROTC) programs at two predominantly black and Chicano high schools in the Los Angeles area. To other persons, especially the higher ranking officers and enlisted personnel who were the managers of the recruiting efforts, affirmative action was a monthly goal for minorities.

The situation was somewhat different in the Marine Corps. Marine Corps people tended to disregard even the need for affirmative action since they did not have an enlisted minority goal. One Marine Corps recruiter commented that he had never given affirmative action any thought because he did not have to obtain minority recruits.

Still another concept of affirmative action which was typical of both services was the placing of minority recruiters in minority areas. To many, this represents "doing something for minorities." Evidence of the Navy's and Marine Corps' recruiters' misunderstanding of the meaning of affirmative action is obvious in the following comments made during field interviews:

> The Navy is looking for quality, regardless of color. There is no need for affirmative action in recruiting because racism is being wiped out in the fleet.
>
> The Marine Corps cannot afford to favor anyone in their entrance requirements.
>
> There is no need for goals in minority enlisted recruiting because the Corps is having no problem with their minority accessions.

This is not to say that these statements are not true or are without merit. In fact, some of them represent what an equal opportunity service should be ideally like. But affirmative action is required in order to reach a point at which equal opportunity is all that is necessary.

RECENT INDICATORS OF RACIAL DISCONTENT

A number of incidents of racial disorder and discontent have occurred in recent years.[33] These clearly indicate the need for the Navy's and Marine Corps' greater implementation of improvements in policies and practices affecting minorities.

---

[33] The accounts of the following two incidents were taken from U.S., Congress, House, Committee on Armed Services, *Report by the Special Subcommittee on Disciplinary Problems in the U.S. Navy*, 92d Cong., 2d sess., January 2, 1973.

U.S.S. Kitty Hawk, *October 1972*

The U.S.S. *Kitty Hawk* is an attack aircraft carrier. On October 10, 1972, the *Kitty Hawk* was in Subic Bay, the Philippines, for supply replenishment and crew rest and recreation. The ship had a crew of 348 officers and 4,135 enlisted men. Five officers and 297 enlisted men were black.

During the evening of the tenth, a fight between blacks and whites occurred in the Subic Bay enlisted men's club. Some of the *Kitty Hawk* black enlisted men were suspected of having been involved in the fighting at the club. On the morning of the eleventh, the ship returned to sea. At approximately 7:00 P.M., the ship's investigating officer called a black sailor to his office for questioning about the Subic Bay fight. The suspect was accompanied by nine other blacks. The group was in a hostile mood and used abusive language while confronting the officer. Only the suspect was allowed into the investigating officer's office for questioning. The suspect was advised of his rights and was released after he refused to make a statement.

Immediately after the investigation, two white cooks were assaulted by a group of blacks which included the suspect released moments before. Within the hour, a large group of blacks congregated on the after enlisted mess deck. The Marine Detachment Reaction Force was called in by an alarmed cook, and a confrontation resulted. The executive officer (XO), a black man, intervened, dismissed the marines, and calmed and dispersed the black sailors.

While the XO was addressing the group of blacks, the commanding officer (CO) entered the after mess deck. The XO was not aware of the CO's presence. (The executive officer is second in command on board a Navy ship. He is junior only to the commanding officer.) Being alarmed by the belligerent attitude of the group, the CO left prior to the dispersion of the group. He ordered the Marine Detachment's commanding officer to post additional aircraft security patrols on the hangar and flight decks. The Marine Detachment's commanding officer in turn gave orders to the marines to break up any group of three or more sailors seen on the flight or hangar decks.

Most of the blacks who had been on the after mess deck with the XO departed via the hangar deck and confronted twenty-six marines. The CO arrived on the scene and placed himself between the marines and the blacks. He ordered the marines to living compartments and attempted to calm the blacks and restore order.

While the CO was on the hangar deck, the XO went below decks where groups consisting of from five to twenty-five blacks were roaming the ship and attacking white sailors. The XO was then informed that the CO had been injured or killed on the hangar deck. Although not sure that this report was true, the XO went to the ship's public address system and ordered the blacks to the after mess deck and the marines to the forecastle in order to separate the two groups.

The CO, neither dead nor injured, was still on the hangar deck talking to the group of black sailors. He was distressed by the XO's announcement. After a brief discussion with the XO, the CO announced over the ship's public address system that the XO had been misinformed and ordered all hands to return to their normal duties. This took place at approximately midnight.

About 150 of the ship's black sailors gravitated toward the forecastle. Their attitude was hostile, and many were armed with chairs, wrenches, and other such weapons. The XO entered the forecastle area and, appealing as one black to another, attempted to control the group. By about 2:30 A.M., the XO was able to persuade the men to put down their weapons and return to their living compartments. This ended the *Kitty Hawk* violence.

Forty-seven men, all but six or seven of them white, were treated for injuries; three men were evacuated to shore hospitals. Twenty-six men were charged with offenses under the Uniform Code of Military Justice. Five were prosecuted aboard the ship during the transit back to the United States. The twenty-one others, after requesting civilian counsel, were put ashore in Subic Bay and flown to San Diego to meet the ship on its return.

U.S.S. Constellation, *November 1972*

The following month, a similar incident took place which further emphasized the Navy's racial unrest. A group of black sailors had been holding meetings between the twentieth and thirtieth of October 1972 on the after mess deck of the aircraft carrier U.S.S. *Constellation*. The group assigned specific functions to its members. Because the blacks were especially intent upon determining whether racial bias had occurred in the awarding of punishment, an examination of the ship's nonjudicial punishment (NJP) records was included in the assignment list.

On November 1, 1972, the CO of the *Constellation* directed the XO to attend the black meeting the following day. General grievances were aired, but no formal complaints were made that could be resolved by internal command action. After the meeting, a group of unidentified blacks assaulted a white cook and fractured his jaw.

The next day, the CO identified fifteen blacks as "agitators." Six were found eligible for administrative discharges in accordance with the guidelines set by the Bureau of Naval Personnel during the Vietnam deescalation to help locate marginal performers and to facilitate their discharge. Processing began immediately, but was later halted.

It was general knowledge at the time that the ship's company would be reduced by 250 men. Room had to be made for air wing personnel soon to be embarked for the upcoming combat deployment. False rumors circulated the ship that all 250 would be black, and that they would be given less than honorable discharges.

On November 3, a sit-in occurred on one of the ship's mess decks. The ship's Human Relations Council (HRC) came in to hear the complaints of the group and met again on the mess decks at 9:00 P.M. the same day for a formal open meeting. The size of the group fluctuated from between 50 and 150, mostly black. By midnight, the group had grown hostile and began hurling verbal abuses at the HRC members, who finally withdrew, leaving the mess decks to approximately 100 sailors. Threatening otherwise to "tear up his ship," the group demanded the CO's presence. The CO, who was on the control bridge conducting flight operations, refused and ordered the ship's berthing and messing spaces patrolled by senior enlisted personnel to avoid the type of violence that had occurred on the *Kitty Hawk*.

After flight operations were completed, the CO informed his seniors that he would return to San Diego and leave the dissident group ashore as a beach detachment. This he did and returned to sea.

On November 8, after the ship had returned to port in San Diego, the CO met with the group to hear its complaints. The group made the following demands: (1) that a review of nonjudicial punishment and administrative discharges be conducted to determine if racial discrimination had occurred against blacks, and (2) that all personnel involved in the incident be allowed back on board and not prosecuted for their actions. The CO agreed to meet the first demand, but deemed that all personnel who had committed prior offenses or who had committed assault during the original incident were to be prosecuted.

The next morning, 122 men from the group mustered on the pier rather than the ship. They were under the misimpression that such a muster would preclude their being charged with an unauthorized absence. The ship's officer of the deck advised them that they were in an Unauthorized Absence Status. At 9:00 A.M., they were subsequently transferred to North Island for disciplinary action. Each man was then charged with six hours' unauthorized absence and fined $25.00.

IMPACT OF INCIDENTS

These incidents rocked the entire military establishment. The U.S. Navy had always boasted that, in its long history, there had never been a mutiny aboard ship. Yet here, within a few weeks of each other, were two incidents which, although not mutinies, were certainly serious breaches of discipline.

The House Armed Services Subcommittee which investigated the incidents cited "permissiveness" in the Navy as a major contributor to the incidents. By permissiveness, the subcommittee meant "an attitude by seniors down the chain of command which *tolerates* the use of individual discretion by juniors in areas in the services which have been strictly controlled; it means . . . a failure to enforce existing orders and regulations which have validity. . . ."[34] The subcommittee also cited the Navy's recruitment in 1972 of a greater percentage of persons in mental category IV and in the lower half of mental category III as a source of "many of the problems the Navy is experiencing today."[35] The subcommittee admonished the Navy for recruiting advertisements which promised more than an individual could reasonably expect from the Navy. This was especially true for those unable to qualify for formal school training.

Admiral Elmo R. Zumwalt, then chief of naval operations, viewed the situation somewhat differently. In a meeting attended by Navy admirals, Marine Corps generals, and Secretary of the Navy John W. Warner after the two incidents, Zumwalt spoke of the failure of the Navy to meet its equal opportunity goals.[36] He did stress the need for greater discipline in this speech and in a later fleetwide message.[37] Zumwalt's chief emphasis, however, was

---

[34] *Ibid.*, pp. 17679-80.

[35] *Ibid.*, p. 17670.

[36] "Zumwalt Rebukes Top Navy Leaders on Racial Unrest," *New York Times*, November 11, 1972, p. 1.

[37] Chief of Naval Operations, "Z-117" (Washington, D.C.: Department of the Navy, November 14, 1972).

on race relations. He was largely concerned with eliminating racial discrimination primarily through the "whole-hearted" implementation of the Navy's equal opportunity program.

The above incidents were not the first of their kind. The proliferation of racial incidents, including the serious racial fighting at Travis Air Force Base, California, and elsewhere, had already prompted the Department of Defense (DOD) to take positive action. Recognizing that minorities in the services were increasing in numbers, and perceiving the need to improve mutual understanding and communication between the races, the DOD ordered the services to provide a minimum of eighteen hours of race relations training per man per year.

The DOD established the Defense Race Relations Institute (DRRI) at Patrick Air Force Base, Florida, in 1971 to train selected career personnel in the armed services to administer and conduct the newly required race relations training. The DRRI provides an intensive seven-week program of instruction which includes such subject matter as minority history and the psychology of prejudice. It also provides effectiveness training for instructors, which includes the study of guided discussion techniques. The school has been attended by officers and enlisted men of many different racial and ethnic backgrounds from all the services. The Navy and Marine Corps acted quickly to complement and to expand upon the DOD race relations training requirements.

NAVY RACE RELATIONS PROGRAM: PHASE I

The Navy instituted Phase I of its race relations program on November 14, 1972. Admiral Zumwalt, then chief of naval operations, approved a plan developed by the Human Resources Development Project Office (Pers-Pc).[38] An accelerated race relations program began immediately.

The Navy Race Relations School (NRRS), presently located at the large training facility in Millington, Tennessee, was established. It provides training, in addition to the DRRI training, for the Navy's Race Relations Education Specialists (RRES). The training is an intensive three-week program during which the prospective RRES are acquainted with general, as well as specific, minority-majority personnel interaction issues. The prospective

---

[38] Bureau of Naval Personnel, Memorandum for the Chief of Naval Personnel on Navy Equal Opportunity Plan, Department of the Navy, Washington, D.C., 1973.

RRES also receive training relevant to the Navy's group-centered race relations seminars.

The race relations program was initially implemented fleetwide by two separate chains of command. The training consisted of two coordinated and complementary educational programs. One was administered by the chief of naval personnel, and the other by the chief of naval training. The former covered all active duty personnel not associated with Navy training commands. The latter provided training for approximately seventy-five thousand personnel undergoing training at the various Navy training commands. In 1974, after it was felt that sufficient servicewide Phase I training coverage was achieved, management of the Phase I program was shifted solely to the chief of naval training, who is now tasked with training newly enlisted and commissioned personnel under Phase I.[39]

Phase I Design

To make all Navy personnel aware of the problem of racism, Phase I is designed to be effective through confronting individuals with examples of racism, both personal and institutional. (Training through confrontation, encounter groups, and so-called T-groups were in vogue during the late 1960s and early 1970s.) Three types of seminars were established and directed at different authority levels. It was hoped that implementation of individual Command Affirmative Action Plans would be achieved as a result of the seminars. Although called affirmative action, the plans were intended to eliminate bias in various command functions, including opportunities for promotion and occupational classification.[40]

As already stated, the Navy's Phase I training was initially accomplished through three types of seminars. It was hoped that, through the Flag Seminars and the Executive Seminars, both increased racial awareness and the development of Command Affirmative Action Plans would take place. The Understanding Personal Worth and Racial Dignity (UPWARD) Seminars were designed to reach a broader base of personnel. All middle and lower level officers and enlisted personnel were required to attend.

---

[39] Bureau of Naval Personnel, Memorandum for the Secretary of the Navy on Navy Human Goals Program, Department of the Navy, Washington, D.C., 1973, Encl. (1), p. 10.

[40] Chief of Naval Operations, OPNAV Instruction 5354.1, Ch. 1, p. 1.

The primary emphasis of the UPWARD Seminars was placed on awareness through confrontation discussions of racial issues.

The Flag Seminars were two-day, twenty-hour seminars. They were designed to help Navy admirals in understanding personal and institutional racism better and to inspire action at the highest level of the Navy's chain of command in order to launch the Navy's affirmative action program. Each Flag Seminar was led by a team of specially trained and qualified RRES acting as facilitators for a group of eight to twelve flag officer participants.[41] The facilitator teams consisted of both white and non-white RRES. In addition, selected minorities were asked to attend each seminar in order to allow for minority input into the otherwise all-white participant group.

The Executive Seminars were three-day, twenty-four hour seminars for personnel in upper management positions.[42] Since upper management varies from command to command in rank and paygrade, it is not easy to characterize the participants on a seniority basis. A large command, such as an aircraft carrier, will have senior personnel in upper management positions, while a small command, such as a minesweeper, will have a very junior upper management. In general, however, a command's Executive Seminar participants were the commanding officer, executive officer, department heads, and key senior enlisted personnel. The Executive Seminars had two purposes: (1) to increase the participants' understanding of personal and institutional racism and (2) to help the participants begin preparing individual Command Affirmative Action Plans. Like the Flag Seminars, the Executive Seminars were led by specially trained and qualified RRES. Because few minorities were of the rank required to be regular participants, a minority input group was selected from the participating command to present their experiences and points of view. This input group was usually a cross section of race and grade.

The UPWARD Seminars were 2½-day, twenty-hour seminars for all personnel in or below middle management positions. The UPWARD Seminars were also led by "salt and pepper" teams. These leaders, although sometimes RRES, were largely volunteers

---

[41] Bureau of Naval Personnel, Memorandum for the Secretary of the Navy on Navy Human Goals Program, Department of the Navy, Washington, D.C., 1973, Encl. (1), pp. 10-11.

[42] *Ibid.*

trained during five-week sessions to conduct UPWARD Seminars in their own commands.

The purposes of the UPWARD Seminars were to increase the recognition of individual worth and to promote understanding and communication between whites and nonwhites. Additionally, the seminar provided a means for its participants to forward to the commanding officer recommendations on issues they collectively found important.[43]

At present, Phase I is being conducted only at officer and enlisted accession points and at formal training commands of duration longer than twenty weeks. The training is conducted in classroom sessions, with the specific format varying according to the distribution of personnel in the particular activity offering the training.[44]

Command Affirmative Action Plans: Concept and Problems

Despite OPNAV Instructions 1500.42 and 5354.1, which required them to produce affirmative action plans, many Navy commands did not develop such plans from their Executive Seminars. One of the possible reasons for this failure at the command level was that Navy units often faced confusion over what an affirmative action plan should address and what it logically could address, given the limited control that individual commands have in implementing such programs. Much of that part of the Executive Seminars addressing awareness of racism dealt with understanding institutional racism. This understanding was facilitated by a minority input group composed of various minority group members from the participating command. Each member of the input group shared his perception of equal opportunity in the Navy and provided accounts of any instances when he personally had felt discriminated against.

The Navy's propensity for standardized testing in recruitment and upgrading and the dearth of minorities in leadership positions were two of the major issues which surfaced in the seminars. When the time came to draft the affirmative action plan, participants realized that the individual command had little or no control over the mechanism which would allow them to address these issues directly. Confusion followed. How could a command take real affirmative action based on the awareness gained in the

---

[43] *Ibid.*

[44] *Ibid.*

seminar when it felt powerless to deal with the major issues? As a result of this confusion, some commands left their seminars with no clear idea of how to begin writing the required action plan.

The purpose of the Command Affirmative Action Plan has been defined as "to insure equal opportunity and treatment and to insure each person in the Navy has true opportunity to rise to as high a level of responsibility as possible, dependent only on the individual's talent and diligence." [45] This conception of affirmative action plans is consistent with Title VII and Executive Order 11246 requirements of nondiscrimination. It does not, however, embody the Executive Order 11246 requirement to take affirmative action to correct problems of minority underutilization. In this sense, Command Affirmative Action Plans, as perceived by the Navy, are actually equal opportunity policy documents. They are not conceived of as instruments to increase minority utilization in areas where minorities are underrepresented.

This concept of the Command Affirmative Action Plan is reasonable considering the Navy's organizational structure. In the Navy and in the Marine Corps, recruiting, promotion, and job placement policies and functions are carried out primarily by large staff organizations independent of regional and local line control. The regional and local line organizations (fleet and shore commands in the Navy and base and afloat commands in the Marine Corps) carry out the primary mission of these services, but have little or no direct control over who is recruited and placed into the various occupational specialties. Nor do they make final promotional decisions. Hence, although line commanders can do much to create a "climate of equal opportunity," they have little control or responsibility for meeting affirmative action goals, as defined in Figures II-1 and II-2. The implications of this circumstance are discussed in later chapters.

Evaluation of Phase I

Despite this shortcoming, Phase I was successful in increasing awareness of racial problems. Participants in UPWARD and Executive Seminars were more aware of unequal opportunity and discrimination in the Navy than were nonseminar attendees. Additionally, seminar attendees experienced increased awareness of individual and institutional racism and had more positive

---

[45] Chief of Naval Operations, OPNAV Instruction 5354.1, Ch. 1, p. 1.

Equal Opportunity and Affirmative Action 27

racial attitudes than had nonseminar attendees.[46] This is truly significant because increased awareness and understanding were two of the major goals of Phase I.

Of seminar attendees, white senior petty officers and junior officers were the least aware of the existence of institutional racism in the Navy.[47] As already stated, Phase I was not successful in generating widespread organizational change and affirmative action. Systems Development Corporation, the company tasked with evaluating Phase I, recommended, "The race relations program should be modified to place increased emphasis on the target population of E7s-E9s and W1s-W4s. . . . To maximize the increased awareness of racial discrimination in the Navy, efforts should be made to incorporate positive corrective actions in operational practices, policies and procedures, e.g., Affirmative Action Plans, formal striker procedures, etc."[48]

The evaluation and recommendations made by the Systems Development Corporation appear sound. The Navy recognizes that the middle management group (junior officers and senior petty officers) was not as effectively involved in the program as it should have been. Furthermore, the Navy realizes that the program failed to create organizational change within the established chain of command. In the fall of 1974, the Navy launched the second phase of its equal opportunity program.

NAVY RACE RELATIONS PROGRAM: PHASE II

The Phase II program is based on consultation rather than confrontation. Phase I used RRES as "experts" who led seminars to confront the attendees with racism in the Navy and their contribution to it. Phase II uses retrained RRES as Equal Opportunity Program Specialists (EOPS) to act as consultants to commanding officers who are implementing equal opportunity requirements.

---

[46] E. E. Arceneaux et al., *Navy Race Relations Education Impact Analysis*, Vol. I (Santa Monica, Calif.: Systems Development Corporation, 1974), p. 2-4.

[47] *Ibid.*, Vol. II, p. 3-12.

[48] *Ibid.*, Vol. I, pp. 3-1 and 3-2. W-1 through W-4 are the four levels in the warrant officer rank. Warrant officers are the lowest ranking officers, falling between ensign and chief petty officer in the Navy's hierarchy. All warrant officers are former enlisted personnel who have moved up through the ranks.

Converting RRES to EOPS

The RRES are required to undergo a conversion training process conducted by the Human Resource Management Centers (HRMC). They are trained to collect and analyze data on equal opportunity, conduct equal opportunity training evaluations, and aid in the creation of affirmative action plans.

As the new name, Equal Opportunity Program Specialists, implies, the emphasis has been shifted from race relations education to equal opportunity implementation. In this regard, the Navy is attempting to involve middle management and produce affirmative action plans relevant to individual commands. The Phase II design calls on each Navy command to collect data on minority representation vis-à-vis majority representation in rates and ratings, nonjudicial punishment, etc. The data collected on each unit are for the edification of the commanding officer. The data are displayed in percentage distribution histograms called Equal Opportunity Quality Indicators (EOQI). EOPS help to prepare and analyze the available data; they also help the commanding officer compare the results with fleetwide command percentages. Finally, the EOPS act as consultants in providing insight into how more effective equal opportunity can be achieved within the command.

Phase II Design

As has been stated, teams of EOPS assist individual units in conducting the Phase II program. The program consists of three interrelated subprograms or tracks: the Affirmative Action Plan Revision Track, the Required Workshops Track, and the Maintenance Track.[49]

The Affirmative Action Plan Revision Track actually begins with the collection of data and the preparation of the EOQI. A Counter Racism/Equal Opportunity Workshop is then conducted. It is a skill-building workshop conducted by the EOPS to provide selected unit personnel with skills useful in developing, updating, and modifying the Command Affirmative Action Plan.[50] An Affirmative Action Plan Development Workshop is

---

[49] *Equal Opportunity Program Specialist Consultant Guide*, Vol. I, Phase II Equal Opportunity/Race Relations Program, NAVPERS 15259 (Washington, D.C.: Department of the Navy, n.d.), p. I-1.

[50] *Ibid.*, pp. V-1 to V-2.

then conducted by the EOPS to provide the personnel trained in the Counter Racism/Equal Opportunity Workshop with the opportunity to refine and update their unit's affirmative action plan.[51]

The Required Workshop Track is composed of three separate workshops. The Middle Management Actions to Counter Racism Workshop is one of the three and is conducted by EOPS for division and warrant officers, chief petty officers, and petty officers leading work groups. Acquainting participants with Navy equal opportunity standards as set forth in the *Equal Opportunity Manual*, the workshop is designed to gain support for the unit's affirmative action plan at the middle management level. The command's equal opportunity situation is assessed by comparing current practices and policies with required Navy standards. Furthermore, the participants develop personal action plans to correct discrepancies at their management levels.[52]

The next workshop in the Required Workshop Track is the Cultural Expression in the Navy Workshop. All personnel are required to attend this workshop. It is conducted by personnel from the individual units who are trained by the EOPS and given the title Command Training Team. The purpose of this workshop is to help participants gain an understanding and tolerance of the different cultural heritages of Navy personnel.[53]

The Military Rights and Responsibilities Workshop is the third required workshop. Also conducted by the Command Training Team, it is mandatory for enlisted personnel in paygrades E-1 through E-4. The purpose of the workshop is to strengthen the Navy's chain of command by increasing the participant's knowledge of the laws, regulations, and customs which the chain of command must enforce. This workshop explores the relationship among rights, responsibilities, and privileges. Furthermore, this workshop trains participants in using the Navy's formal communications channels and formal grievance procedures.[54]

Finally, there is the Maintenance Track, whose purpose is to help the individual unit be able to continue Phase II activities without EOPS assistance. As already mentioned, a Command Training Team is established for each command. The team con-

---

[51] *Ibid.*, p. X-1.

[52] *Ibid.*, p. XII-1.

[53] *Ibid.*, p. XIV-1.

[54] *Ibid.*, p. XIII-2.

sists of individuals from the unit who are trained by the EOPS to conduct the Military Rights and Responsibilities Workshop and the Cultural Expression in the Navy Workshop. A similar group called the Command Information Team is trained to develop, maintain, and update data for the EOQI and to maintain all other Phase II-related information. These two teams are then able to continue the Phase II effort independent of direct EOPS supervision.[55]

MARINE CORPS HUMAN RELATIONS PROGRAM

The Marine Corps has established the Marine Corps Human Relations Instructors Institute in San Diego, California, to train its Human Relations Instructors (HRI). When training its HRI, the Marine Corps uses the institute training either as a supplement or, in some cases, as an alternative to the Defense Department DRRI program. Course material covers various subjects, including philosophy, culture shock problems, minority-majority issues, and a wide range of human relations topics. Much of the institute's emphasis in training is placed on techniques of conducting guided group discussions. The training is done in an intensive 120-hour program. Individuals selected for this program are recommended by their commanding officers and may either be officer or senior enlisted personnel. Marines in grades sergeant through lieutenant colonel are eligible. After completion of training at the Human Relations Instructors Institute, the discussion leaders are then utilized as full-time HRI by their units for at least one year. The Marine Corps meets the race relations requirement through its integrated human relations program, "designed to revitalize within each Marine an understanding of, and an appreciation for, those basic American values necessary for the most effective application of military leadership requirements and principles." [56] The program emphasizes the commonality of man, liberty, and equality as values shared by most men.

Program Design

A regular part of each marine's annual training and qualifications cycle, the program is designed to supplement other Corps-

---

[55] *Ibid.*, pp. I-1 and XV-1.

[56] Commandant of the Marine Corps, Marine Corps Order 5350.4A, p. 5.

Equal Opportunity and Affirmative Action 31

wide leadership training efforts. The training is not concentrated during a two-to-three-day effort. Rather, it is conducted during a series of "guided discussions" over the training-cycle year. In contrast with the Navy's program, the Marine Corps' training effort focuses more on individual development than on group training.

The guided discussion training method used by the Marine Corps consists of a small discussion group and a specially trained discussion leader who uses a detailed discussion manual.[57] The training is conducted over three one-year cycles, each having a corresponding volume of material in the discussion manual which is to be covered. All marines participate in this discussion group training.

Marines in their first year of human relations training enter Cycle I, which places most of its training emphasis on racial problems. Cycle I has three phases.[58] The first phase is a one-to-three-hour orientation, which describes the human relations program and format, emphasizes the program's importance, and acclimates the participants to the guided discussion method.

The second phase consists of a series of guided group discussions, which total between seventeen and nineteen hours of training. Specific topics for discussion include liberty, equality, prejudice, institutional racism, and cross-cultural problems. The discussion leader encourages the discussion group members to express their opinions on the topics and attempts to bring all attendees into the discussion while covering all of the topics in the discussion manual.

The third phase is an action phase in which individual marines are encouraged to take actions which contribute to improved human relations.[59] The discussion leader issues assignments for subsequent discussion sessions. Assignments may entail answering questions raised during a previous discussion or taking some personal action related to the discussion. The emphasis here is on individual behavior modification. Each marine is asked to attempt to alter his behavior to help attain improved human relations in the Marine Corps.

---

[57] Marine Corps Human Relations Institute, *Human Relations Leadership Discussion Manual* (San Diego, Calif.: U.S. Marine Corps, 1973), Vols. I-III.

[58] Commandant of the Marine Corps, Marine Corps Order 5350.4A, p. 2.

[59] *Ibid.*, p. 2.

Cycle II, like Cycle I, has three phases; however, it places less emphasis directly on racial issues. In the first phase, the attendees review the topics covered in the initial twenty hours of training, reflect on their individual action programs, and become reacquainted with the guided discussion method. The second phase is another series of guided discussions, which are conducted in a fashion similar to those in Cycle I. Topics for discussion fall into such categories as cross-cultural adjustment, human-life value, government, war, and male-female relations. The third phase is a continued emphasis on individual action programs. Individual marines are again asked to commit themselves to actions which foster improved human relations.

Finally, Cycle III is an annually repeated cycle for those marines who have completed Cycles I and II. Cycle III is a comprehensive review of the first two cycles and is designed to provide individual unit commanders with the oppportunity to interject human relations and equal opportunity topics of local interest. In addition, the individual action phase is reemphasized.

NAVY AND MARINE CORPS SERVICEWIDE PROGRAMS CONTRASTED

Phase I and Phase II of the Navy's servicewide race relations/equal opportunity program fall into the broader Human Resource Management Program. This broader program is concerned with organizational development, overseas diplomacy, and drug and alcohol abuse education, as well as with equal opportunity. Despite this broader connection, the Navy's equal opportunity program alone imposes servicewide requirements that exceed the DOD's eighteen hours per man per year requirement. The Marine Corps meets this DOD requirement with its servicewide Human Relations Program. As noted earlier, the Marine Corps' program includes material other than that concerned solely with race relations and equal opportunity. It would appear, therefore, that the Navy commits proportionately more man-hours specifically to race relations and equal opportunity than the Marine Corps commits.

Marine Corps unit commanding officers are responsible for ensuring that a trained discussion leader is on board and for ensuring that the men in their commands receive human relations training. The training program itself, however, emphasizes individual behavioral changes and personal actions rather than institutional change and development. The Navy's program, particularly the Executive Seminars of Phase I and the entire Phase II effort, emphasizes individual awareness *and* institutional change, development, and affirmative action planning. It is difficult at this point to assess the relative merits of these two approaches. Certainly the Navy's program is bolder because it recognizes a need for organizational change; but, given the organizational constraints on individual unit commanding officers, the viability of unit affirmative action plans is questionable. Research which assesses the relative strengths and weaknesses of each of these two approaches is greatly needed.

The Navy's race relations/equal opportunity training program is an intensive, one time per year effort when seen from the point of view of the individual sailor. Servicewide Phase I seminars were two-to-three-day sessions; *unitwide* involvement in Phase II normally lasts five days. The Marine Corps' Human Relations Program is a series of guided discussions of approximately one hour each in which individual marines participate over an entire training cycle. Again, the relative merits of these two instructional methods are difficult to assess. Each approach supports the services' respectively different emphasis on individual and institutional change. Still, the contrasting intensity of the two training approaches may have a broader impact on the relative success of the two programs. Again, research is needed to determine the impact of training intensity on program success.

The Navy Affirmative Action Plan of 1976

In November 1975, the chief of naval operations (CNO) formed a task force to review the Navy's progress in the area of equal opportunity and to develop an updated Navy-wide plan for affirmative action. The resulting document, representing a major policy statement by the Navy, was the Navy Affirmative Action Plan (NAAP), issued by the CNO in June 1976. Implementation of the NAAP began immediately.

The plan represents the Navy's desire to institutionalize its efforts toward equal opportunity. The NAAP incorporates ongoing initiatives expressed in other documents dealing with equal opportunity, as well as recommendations and findings of the CNO Task Force. "The format of the NAAP provides broad,

Navy-wide, policy-based direction and guidance to appropriate second echelon commanders, who, in turn, are expected to develop supporting plans for the subordinate commands, organizations and activities under their cognizance." [60]

The NAAP lists specific objectives, affirmative action steps, milestones, and accountable commands. General categories into which objectives are classified include accessions, professional growth, equal opportunity and race relations training, women in the Navy, and assessment and compliance. As stated earlier, second echelon commanders are directed to develop their own supporting affirmative action plans, including any additional areas of affirmative action thought unique to their command.

Another important factor in this plan is the annual review and assessment of the NAAP at all levels within the Navy. "The criteria for progress will be the demonstration of actual, positive movement, trends or changes in the demographics of the Navy's population in terms of ethnic, racial, and sexual composition by rank, rate, career field, retention, disciplinary status and type of discharge." [61]

The NAAP truly seems to be a positive step toward the goal of equal opportunity. The affirmative action steps and review mechanisms in this plan should significantly aid in reaching the objectives contained in the plan. More time must pass, however, before an accurate evaluation of the NAAP can be rendered.

CONCLUDING REMARKS

The Navy and Marine Corps have invested considerable time, effort, and money in pursuit of their race relations and equal opportunity objectives, but recent racial tension at the Marine Corps' Camp Pendleton facility [62] demonstrates that racial harmony has not been achieved. Furthermore, as noted earlier, the Navy is attempting to accomplish its equal opportunity objectives

---

[60] Chief of Naval Operations, *Navy Affirmative Action Plan* (Washington, D.C.: Department of the Navy, 1976), p. 1.

[61] *Ibid.*, p. 4.

[62] See Robert Lindsey, "Uneasy Peace Seen in Marine Camp After Attack on Whites by Blacks," *New York Times*, December 2, 1976, p. 18; Lindsey, "Marines Transfer Klan Leader to Ease Tension at Pendleton," *New York Times*, December 4, 1976, p. 10; and Robert Kaylor, "U.S. Marines on Okinawa Riven by Racial Tension," *Washington Post*, September 9, 1977, p. C5.

through organizational change and affirmative action planning. Similar attempts to use the concepts of organizational development to effect affirmative action have been reported elsewhere.[63]

It is clear that the implementation of an affirmative action program requires more than merely providing a "climate of equal opportunity." The services must continue to identify, analyze, and correct existing barriers to minority entry, upgrading, and mobility. The remaining chapters seek to describe and analyze the services' recruiting, occupational classification, assignment, advancement, and retention policies and their relative impact on minority personnel's upgrading opportunities. Institutional policy modifications based on these analyses are recommended.

---

[63] See Howard E. Mitchell, *An Affirmative Action Program for Two Community Hospitals: A Report of an Organizational Development Program* (Philadelphia: Human Resources Center, The Wharton School, University of Pennsylvania, 1976).

CHAPTER III

Recruitment

The ability of the Navy and Marine Corps to meet their own equal opportunity goals (see chapter II, Figures II-1 and II-2) and to solve their problems with proportionate minority participation depends in large measure on their recruitment organizations, policies, and practices. The discontinuation of the draft and the advent of the all-volunteer armed services has placed the Navy and Marine Corps in direct competition with the private sector for manpower. Much of the private sector, like the services, is seeking to meet minority utilization goals. In spite of this direct competition, the Navy and Marine Corps use organizations and philosophies of manpower procurement that differ markedly from those of most private firms. Private firms' organizations, policies, and practices regarding manpower procurement reflect the competition for manpower. To a large extent, however, the personnel procurement functions of the Navy and Marine Corps continue to operate as if in a draft-motivated environment. The differences in manpower procurement philosophies give private industry an advantage over the services in meeting minority utilization goals.

RECRUITING ORGANIZATION

Figures III-1 and III-2 are summary organization charts which highlight the recruiting organizations for the Navy and Marine Corps, respectively. These manpower procurement organizations are not unlike those found in private firms, especially if allowances are made for the tremendous size of the Navy and Marine Corps.

Like private industry, the Navy and Marine Corps have established recruiting organizations with due regard for area coverage. Like private industry, they give special attention to minority recruiting and provide for the need to use different personnel

FIGURE III-1
Navy Personnel Management and Recruiting Command Organization

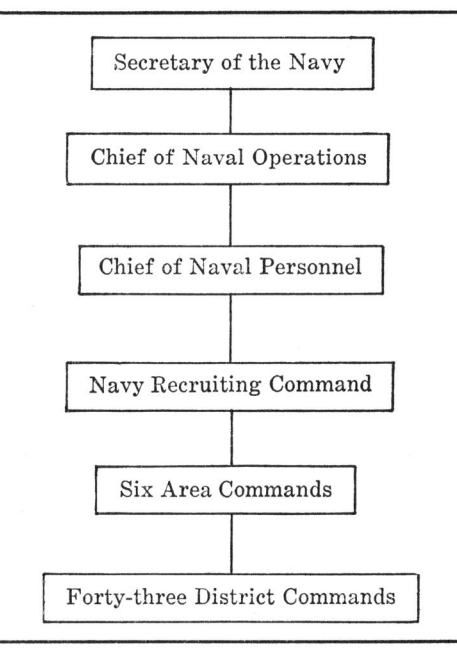

```
            Secretary of the Navy
                    |
            Chief of Naval Operations
                    |
            Chief of Naval Personnel
                    |
            Navy Recruiting Command
                    |
              Six Area Commands
                    |
          Forty-three District Commands
```

Adapted from: Navy Recruiting Command, *Standard Operating Procedures (SOP) Manual*, COMNAVCRUITCOM Instruction 5400.2A (Washington, D.C.: Department of the Navy, 1976).

and techniques with regard to officer and enlisted personnel recruiting. The Navy and Marine Corps, however, assign recruiting to field staffs that operate separately from the line, whereas private industry places that field staff under line control and can thereby hold the line responsible for results.

Line and Staff Relationships

The terms *line* and *staff* have somewhat different meanings in industry than in the military. In industry's terms, the line is composed of the individuals and departments that carry out the primary activities or mission of the organization as a whole. In the Navy and as it is used here, the line is composed of those who direct and man the fleet and bases. In the Marine Corps, it consists of the brigades and companies manning the bases. The staff, on the other hand, is composed of the departments that

FIGURE III-2
Marine Corps Recruiting Command Organization

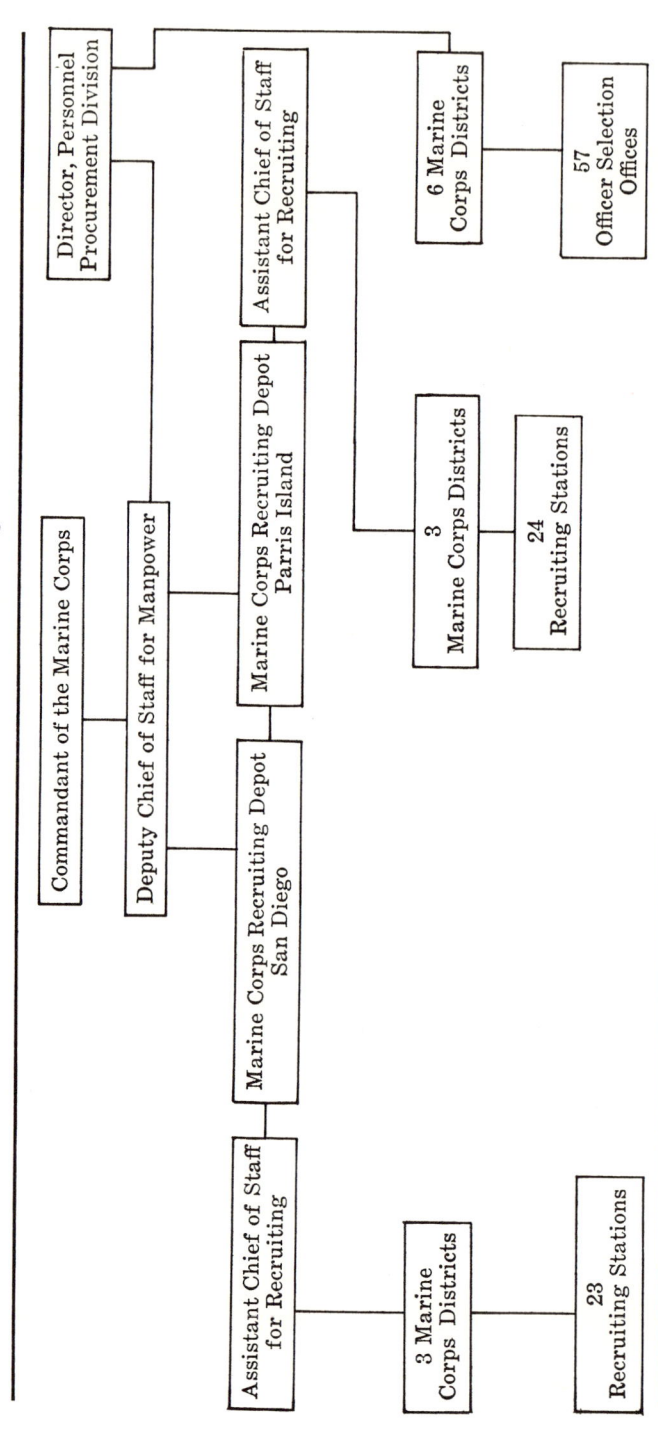

```
                    Commandant of the Marine Corps
                                |
                    Deputy Chief of Staff for Manpower ──── Director, Personnel Procurement Division
                    │                              │
    Marine Corps Recruiting Depot        Marine Corps Recruiting Depot
            San Diego                            Parris Island
                │                                      │
    Assistant Chief of Staff              Assistant Chief of Staff
         for Recruiting                        for Recruiting
                │                                      │
        3 Marine Corps Districts              3 Marine Corps Districts
                │                                      │
         23 Recruiting Stations                24 Recruiting Stations

                        6 Marine Corps Districts
                                │
                        57 Officer Selection Offices
```

Source: U.S. Marine Corps, Headquarters, July 1978.
Note: The Marine Corps Recruiting Depots fall directly under the Deputy Chief of Staff for Manpower for all enlisted operations recruiting matters. The six Marine Corps Districts are directly responsible to the Director, Personnel Procurement Division, concerning recruiting administrative matters and officer procurement.

assist the total efforts of the line departments. In a general sense, staff provides services. Within this definition, the Navy Recruiting Command, for example, is a staff organization, from the admiral in charge to the recruiter in the field.

The organization of the recruiting function in private enterprise is similar to those of the Navy and Marine Corps, but the line-staff relationships are different. Recruiting in private industry is part of the personnel staff function, but responsibility for maintaining and achieving recruiting goals rests with the line. The line plant, department, or division manager is responsible for meeting any goals which may have been set with respect to minority representation and participation in the organization under his control. Thus, the staff function remains one of policy formulation, assistance, and service, and the line function remains one of producing results.

The Navy and Marine Corps have completely dissociated the recruiting function from the operations (i.e., line) function. There is no direct line of responsibility between them and, as a result, no line control. All responsibility for reaching recruiting goals lies with the recruiting staff organization.

In Philadelphia, for example, the Navy and Marine Corps recruiting functions do not report to the chief line officers in that area, but rather report to their respective recruiting organizations. In contrast, General Electric Company, a major Philadelphia employer with at least three large facilities in the area, has a personnel office in each facility, the head of which reports to the plant or department manager there. Recruiting performed by these General Electric personnel managers and their staffs conforms to policies promulgated by the chief personnel officer at company headquarters, but it is done under the direction of plant and department line officers who bear ultimate responsibility for its success.

A centralized manpower procurement system is probably the most effective system for the Navy and Marine Corps. To change from a centralized recruiting system is not considered desirable. Line management, however, has no control over minority recruitment. Therefore, there is no direct line responsibility for minority recruiting goals at the unit command level. As a result, line management is interested in obtaining recruits who can perform successfully, but is not necessarily interested in meeting minority recruiting goals. This is not to imply that the line officers in the Navy and Marine Corps are disinterested, indifferent, or unconcerned about minority participation. It does mean, however, that such concern is not directly linked with their responsibilities.

In private industry, line officials are judged and compensated according to how well they meet the goals necessary for the corporation's success. Affirmative action results are one of these goals. Because line management in the Navy and Marine Corps is not responsible for minority recruitment, there may be little concern for proportionate participation at the command level. There are those who would argue that direct line responsibility for minority recruiting would help to ensure that minority participation goals are met.

THE RECRUITER: SELECTION, TRAINING, AND RESPONSIBILITIES

The selection, training, and responsibilities of the recruiter are key factors in obtaining good recruits. The Navy and Marine Corps are no exception to this principle.

Selection

There are basically two ways in which an individual can be placed on duty as an enlisted recruiter.

1. An individual can request a recruiting post by submitting a formal request to his supervisor. The bulk of Navy recruiters and about one-half of the Marine Corps recruiters are volunteers.
2. An individual can be drafted into a recruiting post. (According to Naval regulations, the individual has the option of rejecting such an assignment without any adverse effects on his career.)

Once a request is submitted, the supervisor (in most cases, the commanding officer of the unit) will examine the candidate's service record and interview the individual to determine whether he would make a good recruiter. All relevant information, along with the officer's recommendation, is then sent to the chief of naval personnel for processing. Once an individual receives the "recruiter rating," he is asked to select three preferences of geographical location. When an opening becomes available in the Navy, the recruiter is assigned to that particular naval area command for a period of three years, although in some cases an extension is given to increase the period to four or five years.

Selection of Navy and Marine Corps recruiting personnel is not influenced significantly by the recruiting establishment. Selection of Marine Corps personnel is done by headquarters. The recruiting organization has no voice in this selection. The Marine Corps recruiting districts determine the number of men needed to carry out their mission, but not the particular men who will be assigned. Navy recruiters are selected on the basis of interviews held with their fleet commanding officers who are neither professional interviewers nor necessarily familiar with recruiting requirements. In the Navy, the Bureau of Naval Personnel screens the volunteers' service record. The Navy Recruiting Command thus lacks complete control over the men who will be assigned as recruiters. Special area considerations, such as the need for effective minority recruiting in some areas, appear to be lacking in the selection process.

There is considerable enthusiasm for the job among new recruiter candidates. For many, it is their first opportunity for shore duty after considerable time at sea. For many others, it represents an opportunity for duty in or near their hometowns that might not be available otherwise. Despite this great enthusiasm and competition for recruiting billets, recruiters are not manpower procurement specialists. They are not selected by the services' recruiting organizations, and selection is not matched carefully to special procurement needs such as minority community penetration.

Training

Before beginning a recruiting assignment, a potential Navy enlisted recruiter must first attend a five-week course in one of the two Enlisted Navy Recruiter Orientation Schools (ENRO). Previously, each of the Navy recruiting areas operated an ENRO school for incoming recruiters, which was staffed by recruiters from the immediate area. ENRO schools were located in the following cities: Albany, New York; Columbus, Ohio; Dallas, Texas; Great Lakes, Illinois; Macon, Georgia; Omaha, Nebraska; Richmond, Virginia; and San Francisco, California. Today, however, ENRO schools exist at only two locations: Orlando, Florida; and San Diego, California.

The objectives of the five-week course (ENRO-I) as developed by the Navy Recruiting Command are as follows:

1. Provide newly assigned recruiters with the basic programming and processing knowledge needed to commence successful field operations as a canvasser recruiter.
2. Provide new recruiters with basic sales skills needed to successfully prospect, and close sales.
3. Acquaint new recruiters with the support functions provided by Commander, Navy Recruiting Command and other supporting elements of the organization.
4. Orient new recruiters in the field of public speaking and community relations, including minority community relationships/school and college relationships.[1]

After the five-week indoctrination, new recruiters are expected to continue training as they carry out the normal duties of recruiting.

ENRO-I schools are manned by senior enlisted men who have extensive recruiting experience and who are responsible for communication of curriculum material to the students. Usually, the only outside lecturer is a representative of a consulting firm who lectures on basic sales techniques.

Once the recruiters have completed this initial training period and have spent time on actual recruiting duty, they become eligible to attend ENRO-II, an advanced training session. These advanced courses are conducted at each of the Navy Recruiting Area Headquarters on an "as-needed" basis. These courses are not meant to be remedial, that is, bring an unproductive recruiter up to par; rather, they are designed to assist the middle manager in his duties as a recruiter supervisor.[2] Time is spent on decision making, trouble shooting, and problem solving.

Under the old ENRO-I curriculum, concentration was on providing new recruiters with a thorough knowledge of Navy programs and applicant-processing procedures (paperwork) rather than with recruiting techniques. Under the new curriculum, a greater amount of time is spent on sales and recruiting techniques and less time on paperwork.

The portion of the ENRO-I curriculum on minority recruiting was underemphasized in the old system and is still not a major area of concentration in the new. Minimal time is spent on the

---

[1] Navy Recruiting Command, *Standard Operating Procedures (SOP) Manual*, COMNAVCRUITCOM Instruction 5400.2A (Washington, D.C.: Department of the Navy, 1976), Section 7218.

[2] *Ibid.*, Section 7217.

specialized recruiting techniques needed in this area. The Marine Corps school involves a seven-week course with much emphasis on paperwork and physical fitness, but no special instruction in minority recruiting. Follow-up training for the Marine Corps is left to the districts. If given at all, such training varies widely.

For officer recruiters, the Navy operates a Recruiting Officer Management Orientation School (ROMO) at Pensacola, Florida, which has a three-week curriculum. Obviously, in such a short period, only a brief treatment of recruiting problems and techniques is possible. Interviews with ROMO instructors and students and with recruiters who had taken the course provided a consensus that more emphasis on interviewing and more instruction on minority-recruiting problems and techniques would be helpful.

The Marine Corps provides a ten-day course at the Xerox Training Center for prospective recruiting station commanders, executive officers, and sergeant majors. Also attending are those officers who will be involved with officer selection/recruiting. This course is designed to cover different aspects of recruiting management. Again, this short time period does not allow for adequate training in race relations.

Job Responsibilities

Once an enlisted recruiter commences his recruiting assignment, he is expected to perform a wide variety of functions in his everyday duty. The Navy and Marine Corps appear to want their enlisted recruiter to be not only a good salesman but also a public relations expert, advertising expert, and market analyst. The recruiter is expected to scrutinize adequately his market for manpower, to prepare the correct advertising approach, to perform public relations efforts aimed at penetrating the manpower market, and, finally, to "sell" that individual within a particular market on the idea of enlisting or being commissioned in the Navy or Marine Corps. He is expected to handle the race relations and minority-recruiting aspects of his job as well. Such a diversity of function presents a challenge for even the most able recruiters. Functional diversity is not as severe for officer recruiters because the program responsibilities for the various officer programs are usually delegated among many officer recruiters.

Length of Assignment to Recruiting Duty

The short tour of duty for most recruiters appears to inhibit overall success. A recruiter normally spends the first six months or more finding out what the actual recruiting job demands. After he has established a good relationship with the community in which he works and learns how to tap that particular manpower market, it is often then time for him to return to his regular military occupation. The long process needed to establish credibility in the minority community is often not possible within these short-duty tours.

This is a difficult problem for the Navy and Marine Corps. Men must be rotated off ships and from far-off duty stations. To lengthen the recruiting tour would mean that some men would have to wait longer for shore duty or duty in the continental United States. The problem of duty tours which are too short will continue to make the task of recruiting minorities for the Navy and Marine Corps difficult.

In March 1978, the Navy announced the establishment of a Career Recruiter Force (CRF).[3] This force will eventually consist of about 750 people or about 25 percent of the total Navy recruiting force (300 people are to be selected in fiscal 1978). These people will fill key recruiting management positions and recruiter/canvasser billets. Eligible are those in paygrades E-5 through E-9 and who have had at least two years of recruiting duty since 1973. People selected for the CRF will be identified with an NEC.[4] E-5 and E-6 personnel, however, will continue their normal ship/shore rotation. While assigned to ships, they will continue in their primary occupations and, when rotated ashore, will be assigned recruiter duty.

There are two advantages of a professional specialized recruiting organization. First, the recruiting specialists can develop and maintain an effective recruiting methodology based upon current manpower procurement research. Effective recruiting techniques would then be taught to new recruiters by professionals. Second, a consistent framework for decision making can be established by the recruiting organization. A professional, specialized organization could then make decisions and implement

---

[3] Chief of Naval Operations, "Career Recruiter Force" (Washington, D.C.: Department of the Navy, March 31, 1978).

[4] NEC is Navy Enlisted Classification, a numerical designator for jobs within the Navy.

policy based upon experience and experimentation in the manpower field.

Prior to the establishment of the CRF, the Navy made some attempts to professionalize the recruiter billet through the use of the Navy counselor rating. This rating, with an E-6 entry level and still serving an important function, is an attempt to establish a professional base of recruiters who, while on duty as recruiters, will function as both specialists in the field of recruiting and trainers of other recruiters. The latter is an especially important aspect of their duties because the Navy Recruiting Command intends that these counselors be employed as zone supervisors or chief recruiters who oversee the on-the-job training of new recruiters.

Once a Navy counselor finishes his tour of duty as a recruiter, the Navy Recruiting Command conceives of his next duty as that of career counselor. Essentially, he will be in the retention business, i.e., counseling persons on reenlistment opportunities, etc. It is hoped that these individuals will eventually return to recruiting duty, but this may not necessarily be the case.

RECRUITING FACILITIES AND FACILITY MANNING

Recruiting stations are usually the initial contact point between a recruiter and a person seeking information on entrance qualifications for service as a Navy or Marine enlisted person. The location and appearance of the facility, as well as the impression made by the personnel manning that station, can have a positive or negative effect on potential recruits. This is especially significant for minority recruiting.

Location

The need for flexibility and mobility is very great in minority recruiting. Location of an office in one ethnic or racially dominated area can greatly hinder access by other ethnic or racial groups. One's "turf" in the city is significant and can determine whether minorities are attracted to, or repelled from, the recruiting station.

Considerations for the proposed location of a station depend, to a large degree, on the Qualified Military Available Statistics, which ostensibly indicate the availability of manpower. Little consideration is given to economic conditions, sociological conditions (gang warfare, etc.), or the station's accessibility to the potential market. The offices are generally stationary, but the

Navy has taken some steps to alleviate the immobility factor by assigning mobile vans to various recruiting districts.

Management at the district level of the Marine Corps and Navy recruiting structures, although away from the actual recruiting location, plays an important role in the determination of a particular recruiting office's location. This situation, along with the sometimes difficult task of obtaining a General Services Administration approval for a requested location change, appears to cause the recruiting organization to respond slowly when a recruiting station ought to be moved because of changing economic patterns, educational levels, or age distribution of the market in which it will recruit. Minority recruiting needs, per se, are not significantly considered in the location decision.

Minority Community Penetration Programs

The Navy and Marine Corps must make a concerted effort to penetrate minority communities. By doing this, the services can increase their chances of (1) *identifying* those minorities who already possess the qualifications for a successful military career and (2) *identifying and grooming* promising minorities who are presently not qualified for service because of deficiencies in "academic credentials." It is clear, however, that the key to the services' ability to improve minority recruitment is the penetration of minority communities.

Frank Cassell, former director of the United States Employment Service, points out that the effective penetration of depressed urban areas "means going into the pool halls, bars, street corners, alleys and other places where the unemployed in the slum areas hang out. . . ."[5] He further points out that the most effective way to conduct such penetration programs is to use trained people who are indigenous to such neighborhoods.

The Navy has begun to experiment with the development of penetration programs.[6] Since 1974, the Urban League of Philadelphia, under contract with the Navy, has been providing assistance to the Navy in minority recruiting in urban Philadelphia.

---

[5] Frank H. Cassell, "Creating Human Assets—Not Liabilities" (Paper presented to the Washington Workshop Seminar on Labor-Management Relations sponsored by American Society for Personnel Administration and the Bureau of National Affairs, Inc., Washington, D.C., March 16, 1967), p. 10.

[6] Information concerning the Navy contracts with the Urban League was obtained from Navy recruiting command officers in Philadelphia and Washington, D.C. Both officers were directly involved with contract negotiation and/or implementation.

Recruitment

The approach being taken is to identify *qualified* minorities and encourage them to join the Navy.

The Urban League's efforts are concentrated in three programs. The first program calls for the development of "outreach centers." Outreach centers are actually literature racks containing Navy recruiting information at key points in youth facilities (such as the YMCA). The second program provides for a series of seminars with school guidance counselors. This effort is designed to inform school administrators of the opportunities available in the Navy for qualified high school graduates. Activities include visits to Navy facilities, movies, lectures, informal group sessions, etc. The third program is a seminar series for urban youth. The purpose of this effort is twofold. First, the program is designed to expose high school juniors to the various college testing programs such as the SAT or ACT. Participants are introduced to testing methodology and general content and are given instructions for preparation for such tests. Second, information is disseminated concerning Navy officer programs. It is hoped that such efforts will in fact provide for increased awareness of, and interest in, Navy careers. It is likely that programs of this nature will require an expanded program of remedial education for those having the capability but lacking the educational experience.

Facility Manning for Recruiting Stations

Discussions with recruiting organization personnel indicate that manpower for a recruiting station is primarily a function of the Qualified Military Available Statistics for a particular area. Other indicators used are the station's history of productivity and the number of high schools in the area.

Navy and Marine Corps recruiting managements are reluctant to assign recruiters by race or ethnic origin to particular stations. Their attitudes reflect the belief that the equal opportunity practiced within the Navy as the "One Navy" concept and within the Marine Corps as "All Marines are Green" should be practiced in their recruiting policies. This implies that a good recruiter should be able to recruit anyone of any race or ethnic group. Those who actually man the recruiting stations, however, feel, more often than not, that a black recruiter recruits better in a black community, that a Chicano recruiter recruits better in a Chicano community, and that this holds true for other racial and ethnic groups as well. It is probably true that the race-ethnic characteristics of a recruiter make a difference in recruitment performance. Although not ideally desirable, the race-ethnic characteristics of a recruiter should probably be taken into consideration in the short run when determining regional assignment. Although such assignments should be made with care, an attempt to assign personnel to selected stations while considering race-ethnic information might increase minority accessions. It is hoped that, in the long run, the need for this consideration will not exist, and that the "One Navy" and "All Marines are Green" ideals will be realities.

MENTAL TESTING: THE KEY TO ENLISTED RECRUITING

The keystone of the services' enlisted recruiting function is mental testing. Written examinations determine eligibility for service entry, whether the entrant receives specialized training, and what type of training the recruit is eligible to receive.

Recruiting/Training Relationship

Inherent in any system of recruitment, whether voluntary or compulsory, is a set of standards which determine which persons shall be eligible for selection. The classification "4F—mentally, morally, or physically unfit" used during World War II and thereafter was a designation that, by certain criteria, some persons were unfit for military duty. This or any other standard requires definition by designated criteria. In turn, the criteria must be shaped by the needs of the armed services, the requirements of public policy, and the realities of the labor market.

The armed forces have had an extremely high turnover rate with which few, if any, private employers must deal. Because of this high turnover rate, the services have been interested in recruits who can be trained rapidly and can spend the bulk of their active service with operational units.

All prospective employers attempt to judge and categorize potential recruits. The Navy and Marine Corps do so to a much greater extent than most private sector firms. In a very real sense, for the Navy and Marine Corps, *recruiting is a function of training*. The services attempt, primarily on the basis of written tests, to decide what capabilities a person must possess to be trained in a given occupation. Then, the services try to recruit personnel that meet these test-established qualifications.

Given minorities' disadvantages in education, work experience, and test exposure, it is unlikely that the Navy and Marine Corps can solve their problems with recruiting minorities if they adhere to this philosophy.

Mental Group Determination

The use of mental standards in the armed services dates from the passage of the Selective Service Act of 1948. Potential recruits are classified into five (actually seven, since two are subdivided) mental group categories. I is the highest, V the lowest.

The criterion used to determine an individual's mental category is that individual's score on the Armed Forces Qualifying Test (AFQT). (All of the services are now using the Armed Services Vocational Aptitude Battery [ASVAB]. See chapter V for a complete discussion of testing and AFQT/ASVAB score conversion.) Table III-1 shows the minimum AFQT percentile score required for each mental category. Mental standards have varied over the years subject to the manpower requirements of the services. Thus, mental group standards tend to be somewhat arbitrary. At present, individuals in mental groups lower than IV and V are not eligible for enlistment.[7] Some individuals so classified are, however, still found in the services, a result of less stringent recruiting policies in earlier periods.

In addition to qualifying mentally, prospective recruits must also qualify physically and morally. The mental qualification is, however, the first and most pervasive criterion for service entry. Before being screened physically and morally, an applicant must achieve an AFQT percentile score of 21 or higher. No waivers are allowed for scores below 21.

Moral and Physical Criteria

Applicants who achieve a sufficiently high AFQT score then have their records screened to determine their moral qualifications. Arrest records and admissions of previous drug use are among the most common moral disqualifiers. Depending upon the nature of the offense, however, waivers may be granted at the district, area, or headquarters level of the recruiting organization.

---

[7] Assistant Chief of Personnel Planning and Programming, "Recruiting Goals and Policies," Memorandum for Navy Recruiting Command, Washington, D.C., September 22, 1977, p. 1.

TABLE III-1
Mental Group Classification and AFQT Score

AFQT Score	Mental Group
93-99	I
65-92	II
49-64	Upper III
31-48	Lower III
21-30	Upper IV
10-20	Lower IV
20 \	V

Source: Navy Recruiting Command, *Navy Recruiting Manual—Enlisted*, COMNAVCRUITCOM Instruction 1130.8A (Washington, D.C.: Department of the Navy, 1975), p. 4-4.
Note: AFQT percentile score is derived from the ASVAB testing results.

It is important to note that requests for waivers of moral disqualifiers are initiated by recruiters in the local recruiting office. Such requests are not automatic. The individual recruiter uses his own discretion in determining whether a waiver request will be submitted for higher level consideration. Once submitted, each request is judged on its own merit. It is quite possible for a recruiter to *request* such a waiver for one individual and not for another individual with similar qualifications and a similar offense. It is also possible for a waiver request to be *granted* to one individual and not to another with similar qualifications and a similar offense. The potential for racial bias is obvious.

Service applicants must also pass a comprehensive physical examination. Physical standards vary from service to service. For the Navy and Marine Corps, the Bureau of Naval Medicine has the authority to grant waivers for certain physical defects which may otherwise disqualify an applicant. Again, such waivers are granted on an individual basis, and the potential for racial bias exists.

The AFQT score is the most rigid and pervasive of the service eligibility requirements. Not only does it determine initially whether an applicant will be further screened for moral and physical qualifications, but it is also used in determining whether waivers will be granted. A high enough score can offset other less desirable qualities, while a low enough score can eliminate an otherwise desirable candidate.

Recruiting Mix

The Navy currently uses the dual criteria of school eligibility and high school graduation in determining its recruiting mix. To be "school eligible," [8] a prospective recruit must achieve one of the following:

1. an AFQT percentile score of 50 or higher;
2. an ASVAB composite score of WK plus AR of 100 or higher; or
3. enlistment with a class A school guarantee.[9]

Figure III-3 illustrates the combination of high school graduation and school eligibility used in determining recruiting mix.

FIGURE III-3
Navy Recruiting Criteria

	High School Graduate	Non-High School Graduate
School Eligible	A	B
Non-School Eligible	C	D

Source: Area Level Navy Recruiting Command (Philadelphia), June 1976.

Applicants in category A are both school eligible and have high school diplomas; those in category B are school eligible, but non-high school graduates; those in category C are high school graduates who are not school eligible; and those in category D are neither high school graduates nor school eligible.

The proportion of people enlisted from each category is based on the needs of the Navy and changes over time. An example would be as follows: Individuals in categories A, B, and C may be enlisted at a ratio of five A and/or B per one C, while individuals in category D may not be enlisted. The latter statement concerning category D personnel is always true, as is the fact that the General Educational Development (GED) high school

---

[8] "School eligible" is a quality measure and does not necessarily guarantee that a recruit so designated will be sent to A school. See Navy Recruiting Command, *Navy Recruiting Manual—Enlisted*, COMNAVCRUITCOM Instruction 1130.8A (Washington, D.C.: Department of the Navy, 1974) for a complete discussion of the "school eligible" quality measure.

[9] Assistant Chief of Personnel Planning and Programming, "Recruiting Goals and Policies," p. 1.

equivalency diploma, awarded to non-high school graduates upon successful completion of an educational achievement examination, is not accepted for individuals with AFQT scores between 21 and 30 inclusive.[10]

The Marine Corps' recruiting mix is based largely on educational attainment (i.e., high school graduation). In passing the fiscal 1974 Defense Appropriations Bill, Congress was concerned with the decline of service recruits with high school diplomas. As a result, it ordered each service to recruit no more than 45 percent non-high school graduates in fiscal 1974.[11] The Marine Corps was the only service unable to meet this requirement; it recruited 53 percent high school graduates.[12] The Marine Corps did somewhat better in fiscal 1975 by recruiting approximately 59 percent high school graduates.[13]

The Senate Armed Services Committee, in its report on the fiscal 1976 Defense Appropriations Bill, requested that the commandant of the Marine Corps undertake a study of the Marine Corps' manpower and quality levels, mission, and force structure for submission by January 1, 1976.[14] One of the results of this study was the setting of recruiting goals to attain 67 percent high school graduates for fiscal 1976 and 75 percent for fiscal 1977. During the first five months of fiscal 1976, male non-prior service accessions included 73 percent high school graduates.[15] Thus, the Marine Corps' recruit qualification mix is determined by high school diplomas and minimum AFQT scores of 21.

Academic credentials are the most pervasive variable used in the recruitment process. The moral and physical attributes of an individual are considered only as potential disqualifiers. There is no explicit positive consideration given for individual, nonacademic virtues or for past nonacademic achievement. Thus, an individual may be disqualified from further recruitment consideration because of high school expulsions or an arrest record.

---

[10] Navy Recruiting Command, COMNAVCRUITCOM Instruction 1130.8A, p. 4-4.

[11] Martin Binkin and Jeffrey Record, *Where Does the Marine Corps Go from Here?* (Washington, D.C.: The Brookings Institution, 1976), p. 60.

[12] *Ibid.*

[13] *Ibid.*, p. 61.

[14] See U.S. Marine Corps, Headquarters, "Report on Marine Corps Manpower Quality and Force Structure" (Washington, D.C., 1975).

[15] *Ibid.*, pp. 2, 22.

Recruiting policy does not formally weight such positive achievements as high school student government experience, varsity athletic letter achievement, or Boy Scout rank.

As is discussed in chapter V, the same mental tests used for service entry are also used to determine which of the new recruits will receive formal skill training and for which occupational categories a recruit is eligible. Hence, mental testing, the keystone of the services' recruiting programs, is used as a determinant of who is recruitable and who is trainable and as a limiting criterion in meeting recruiting mix objectives. The impact of standardized testing on minority personnel is discussed in the next section.

SERVICE ENTRANCE STANDARDS AND MINORITY ACCESSIONS

Minorities typically have lower aptitude test scores than whites.[16] Many reasons have been offered for these test score variances: quality of education, lack of familiarity with standardized test taking methods, and language or other communications problems.[17] In the services, these lower aptitude scores are reflected in the heavy concentration of minorities in the lower mental group classifications. Therefore, increasing the minimum acceptable AFQT cut-off score for recruitment affects minority personnel more severely than majority personnel.

Tables III-2 and III-3 show the seriousness of this problem for the Navy and Marine Corps, respectively. They list the number and proportion of enlisted accessions by race and mental group for 1977. It should be noted that personnel classified as mental group V are no longer enlisted in the services; therefore, the data presented are for mental groups I through IV only. The tables

---

[16] John Garcia, "I.Q.: The Conspiracy," *Psychology Today*, Vol. 6, No. 4 (1972), pp. 40-43. This article is part of a series concerning I.Q. testing. The AFQT aptitude examination is, to a large extent, an intelligence test. It is common knowledge that, in the aggregate, minority personnel do not score as high on these so-called intelligence tests. Garcia believes that these types of exams should be used only for the purpose of measuring scholastic achievements. The article denies innate differences in intelligence across race-ethnic groups. Research indicates that similar socioeconomic backgrounds provide for similar intelligence test scores across race-ethnic groups.

[17] For a summary of research on group test score variances, see Arthur I. Siegel, Brian A. Bergman, and Joseph Lambert, *Nonverbal and Culture Fair Performance Prediction Procedures*, Vol. II, *Initial Validation* (Wayne, Pa.: Applied Psychological Services, Inc., 1973), pp. 2-7.

TABLE III-2
Navy Enlisted Accessions by Race and Mental Group
for Fiscal 1977

Mental Group	Caucasian	Percentage Caucasian	Black	Percentage Black	Other Minority	Percentage Other	Total	Total Percentage
I-IV	86,931	85.8	10,890	10.8	3,452	3.4	101,273	100.0
I	7,053	97.0	111	1.5	108	1.5	7,272	100.0
II	30,628	93.7	1,479	4.5	590	1.8	32,697	100.0
III	47,535	80.7	8,750	14.8	2,601	4.4	58,886	99.9 [b]
IV [a]	1,715	70.9	550	22.7	153	6.3	2,418	99.9 [b]

Source: Bureau of Naval Personnel and Naval Recruiting Command, "Chief of Naval Recruiting Command (CNRC) Program Analysis" (Washington, D.C.: Department of the Navy, 1978).

[a] IV refers to mental group upper IV only.
[b] Percentages subject to rounding errors.

TABLE III-3
Marine Corps Enlisted Accessions by Race and Mental Group for Fiscal 1977

Mental Group	Caucasian	Percentage Caucasian	Black	Percentage Black	Other Minority	Percentage Other	Unknown	Total	Total Percentage
Total	33,544	74.5	9,465	21.0	1,886	4.2	153	45,048	99.7 [a]
I	1,891	96.7	42	2.1	23	1.2	—	1,956	100.0
II	10,928	89.8	991	8.1	256	2.1	1	12,176	100.0
III	19,922	69.1	7,434	25.8	1,457	5.1	1	28,814	100.0
IV	800	41.2	993	51.1	150	7.7	—	1,943	100.0
Unknown	3	N.A.	5	N.A.	—	N.A.	151	159	N.A.

Source: U.S. Marine Corps, Headquarters, Manpower Planning, Programs and Budgeting Branch, August 1978.
[a] Unknowns not included in total percent.

show that minorities in both the Navy and the Marine Corps are overrepresented in mental group IV.

Table III-4 shows a comparative percentage distribution of enlisted accessions by race and mental groups for the Navy and Marine Corps and indicates the concentration of minorities in mental groups III and IV. For the Navy, over 43.0 percent of Caucasian enlisted personnel, but only 14.6 percent of blacks and 20.2 percent of other minorities, are in mental groups I and II. On the other hand, the proportion of blacks in mental group IV is more than twice the proportion of Caucasians in that mental group. For the Marine Corps, over 38.0 percent of Caucasian personnel, but only 11.0 percent of blacks and 15.0 percent of other minorities, are in mental groups I and II, whereas 2.4 percent of Caucasians, 10.5 percent of blacks, and 7.9 percent of other minorities are in mental group IV.

Mental group IV personnel are not eligible for specialized formal school training. Those enlisted personnel categorized as mental group III (80.3 percent of blacks and 75.4 percent of other minorities in the Navy and 78.6 percent of blacks and 77.3 percent of other minorities in the Marine Corps) are considered as having potential for limited training.

In the past, neither mental group IV personnel nor mental group III personnel have qualified for the more technical or skilled occupational training. A large percentage of the minorities in the Navy and the Marine Corps are classified as mental group III and, therefore, are eligible only for less technical formal school training. Additionally, greater than 5.0 percent of the black accessions in the Navy and over 10.5 percent of the black accessions in the Marine Corps entered the military classified as ineligible for specialized formal school training. Individuals not receiving formal school training are at a disadvantage in promotion rate relative to individuals with specialized training. Obviously, the outlook for significant improvement in the job and rank distribution of minorities by training and upgrading is limited, at least in the short term, by the mental testing criterion imposed by the services.

The Impact of the Educational Criterion

High school graduation is the other major recruit-screening criterion used by the services. The Navy's recruiting goals for fiscal years 1977 and 1978 called for 82 percent high school graduates, of which 76 percent would be bona fide diploma hold-

Recruitment

TABLE III-4
Percentage Distribution of Navy and Marine Corps Accessions by Race and Mental Group for Fiscal 1977

Mental Group	Navy Caucasian	Navy Black	Navy Other	Marine Corps Caucasian	Marine Corps Black	Marine Corps Other
I	8.1	1.0	3.1	5.6	0.4	1.2
II	35.2	13.6	17.1	32.6	10.5	13.6
III	54.7	80.3	75.4	59.4	78.6	77.3
IV	2.0	5.1	4.4	2.4	10.5	7.9
TOTAL	100.0	100.0	100.0	100.0	100.0	100.0

Source: Derived from Tables III-2 and III-3.

ers and the remainder would hold GED diplomas.[18] The Marine Corps, as noted earlier, established the requirement to recruit 75 percent high school graduates in fiscal 1977.[19] Additionally, Congress has established 55 percent as the minimum percentage of high school graduates that any of the services may recruit.

Like mental testing, this criterion falls more heavily on minorities than on whites. In 1974, 78.8 percent of white males between the ages of eighteen and twenty-four had completed or gone beyond high school. For minority males in the same age group, only 60.2 percent of blacks and 50.2 percent of the Spanish heritage group had completed or gone beyond high school.[20]

Educational and Mental Standards, Minority Accessions, and the Economy

Table III-5 demonstrates the extent to which increased quality standards affect minority accessions in the Navy. The table shows black accessions as a percentage of total monthly enlisted accessions for nine months of fiscal 1975. The table also shows the monthly unemployment rate as a percentage of the total work force over the same period.

---

[18] Assistant Chief of Personnel Planning and Programming, "Recruiting Goals and Policies," p. 2.

[19] U.S. Marine Corps, Headquarters, "Report on Marine Corps Manpower Quality," p. 2.

[20] U.S., Department of Commerce, Bureau of the Census, *Current Population Reports*, Series P-20, No. 274, "Educational Attainment in the United States: March 1973 and 1974," pp. 17, 18, 20.

The percentage of blacks recruited decreased substantially between December 1974 and February 1975. In December 1974, the Navy attempted to upgrade the quality of its new recruits by accepting fewer applicants in mental group IV.

The increase in mental aptitude recruitment standards occurred at a time when the national economy was reaching the depths of the 1974 recession. Unemployment was high and increasing. As unemployment increased and job alternatives became scarce, there were, perhaps, a greater number of more highly qualified persons, generally white, willing to join the Navy in order to find secure employment, and the Navy was enabled to meet its recruitment goals.

Black accessions may thus be viewed as a function of the unemployment rate. Simple regression analysis with black accessions as the dependent variable and the unemployment rate as the independent variable was performed on the data in Table III-5. Eighty-nine percent of the variance in the percentage of monthly black Navy accessions was accounted for by changes in the monthly unemployment rate. Such a high regression coefficient indicates some relationship between minority recruiting and the state of the national economy.

Quality Impact

Despite the apparent ill effects of increasing quality standards on minority accessions, the services point out that the result is not only improved overall force quality but also greater minority upward mobility as well. The Marine Corps claims that the "mere possession of a high school diploma is statistically the most reliable preenlistment predictor that an individual will perform successfully from recruit training through expiration of active service. . . ." [21]

Military manpower specialist Edward Scarborough noted with concern the drastic reduction in minority enlistments in all of the services, especially between July 1974 and March 1975 (see Table III-5). During this period, the number of blacks inducted into the Navy fell from 15.5 percent of the total number of recruits to 5.9 percent. The Marine Corps enlistment of blacks fell from 22.0 percent to 19.0 percent. The Pentagon states that, as of the end of 1975, approximately 16.0 percent of the armed

---

[21] U.S. Marine Corps, Headquarters, "Report on Marine Corps Manpower Quality," p. 5.

TABLE III-5
Navy and Marine Corps Black Enlisted Accessions as
a Percentage of Total Accessions: Unemployment Rate as
a Percentage of Total Civilian Work Force

Month	Navy Black Accessions	Marine Corps Black Accessions	Unemployment Rate
7/74	15.5	22.0	5.3
8/74	14.7	22.0	5.4
9/74	12.7	23.0	5.8
10/74	12.0	24.0	6.0
11/74	11.6	25.0	6.5
12/74	9.7	22.0	7.1
1/75	6.4	21.0	8.2
2/75	5.8	20.0	8.2
3/75	5.9	19.0	8.7

Sources: Column 2—Bureau of Naval Personnel (Pers-611), Washington, D.C., June 1978; column 3—Edward Scarborough, "Minority Participation in the DOD" (Washington, D.C.: Defense Manpower Commission, 1976); column 4—U.S., Department of Labor, Bureau of Labor Statistics, *Employment and Earnings*, Vol. 21. Nos. 7 and 12, p. 54.

forces was black. Today, approximately 12.0 percent of the United States military-age population is black. Approximately 18.5 percent of the Marine Corps and 8.0 percent of Navy enlisted personnel are black.

Scarborough claims that the armed services discriminated against potential black recruits in order to maintain a "racial balance." He indicates that recruiters were given a ceiling on the percentage of blacks that could be recruited. Scarborough further states that it was apparent during the period 1974 to 1975 that the "management or regulation of black accessions (enlistments) was actively pursued by the services without legally approved Department of Defense, Congressional or public policy." [22]

A second study, prepared jointly by Kenneth J. Coffey and Frederick J. Reeg, points to recruiting malpractices and racial

---

[22] Edward Scarborough, "Minority Participation in the DOD" (Washington, D.C.: Defense Manpower Commission, 1976), p. 15.

discrimination practices of the armed services. The study claims that recruiters were ordered to turn away many blacks regardless of their qualifications and, in so doing, sometimes had to accept unqualified whites to meet recruiting quotas.[23] This contention appears to be supported by the testimony of two former Marine Corps recruiters who were questioned by a Senate Armed Services subcommittee investigating recruiting malpractices in the Marine Corps. Each of the former Marine recruiters claimed that he was limited to a maximum quota on the number of blacks that could be enlisted in the Marine Corps.[24]

Deputy Assistant Secretary of Defense H. Minton Francis partially confirmed the above allegations. He stated that recruitment of black men was restricted by the services between 1973 and 1975. He further contended that the services adopted policies such as increased AFQT cut-off requirements that effectively kept blacks out of the military in large numbers, although minority control was not the stated objective. According to Francis, the Defense Department feared that the all-volunteer armed services, instituted in 1973, would eventually become predominantly black.[25] He also stated, "It was the spreading of a paranoia that went rapidly through the management of the four military services."[26] When these allegations became public in June 1976, the Army, Air Force, and Marine Corps denied that they sought to curtail the number of black recruits, and the Navy made no immediate comment on the subject.[27]

PROJECT 100,000: A CASE OF LOWER MENTAL STANDARDS

Through Project 100,000, the services provided a means by which individuals scoring below the usual mental standards cut-off point could enlist in the armed forces. This experiment

---

[23] Lionel Bascom, "Unwelcome in the Military," *Philadelphia Inquirer*, June 17, 1976, p. 1-A.

[24] "2 Assert Marines Had Racial Quota," *New York Times*, June 6, 1976, p. 37.

[25] Bascom, "Unwelcome in the Military," p. 4-A.

[26] *Ibid.*

[27] "Military Accused of Bias in Recruiting," *New York Times*, June 9, 1976, p. 10.

provides at least a crude measure of how well people with low mental aptitudes perform in the services relative to those who have qualified under the standard mental aptitude criterion.

Background

In 1965, then Secretary of Defense Robert S. McNamara became concerned over the high draft rejection rate. Many of the men being rejected had failed to meet the services' test-based mental standards. McNamara became "convinced that at least 100,000 men a year who were being rejected for military service, including tens of thousands of volunteers, could be accepted." [28] In August 1966, he announced his intention to accept 100,000 men per year, beginning in fiscal 1967, who would not have previously been enlisted because of low scores on the mental aptitude test.[29]

The new policy allowed high school graduates with a minimum AFQT percentile score of 10 to enter the services without taking supplementary aptitude tests. Also accepted were non-high school graduates scoring 21 or higher on the AFQT. Non-high school graduates scoring between 10 and 21 on the AFQT might be enlisted pending the outcome of supplementary aptitude tests.[30]

Evaluations of Project 100,000

Initial evaluations of Project 100,000 were optimistic. During the first year of the program, 96 percent of the men who entered the service under the new standards, compared with 98 percent of other new recruits, successfully completed recruit training.

Later evaluations, however, were less encouraging. In June 1973, some Navy findings concerning Project 100,000 were summarized as follows:

> 1. In spite of the label "untrainables" attached to them, Category IV men can be trained to serve in a number of skills, although they require considerably longer training periods than other Navy personnel.

---

[28] Robert S. McNamara, *The Essence of Security* (New York: Harper & Row, 1968), pp. 131-32.

[29] Harold Wool, *The Military Specialist* (Baltimore, Md.: The Johns Hopkins Press, 1968), p. 181.

[30] *Ibid.*, p. 182.

2. Category IV men experienced more disciplinary actions, completed fewer training courses, required more supervision on the job, and advanced in paygrade at a slower rate than did Non-Category IV men.

3. Retention of Category IV men is much lower than that of other Navy men; very few of these men serve over three years and most of the discharges are at the convenience of the Navy.[31]

These findings were meaningful even though, in 1972, mental standards were raised, thus effectively discontinuing Project 100,000.

Entrance requirements have fluctuated over time as a result of the combined effects of military manpower requirements and labor market conditions. Therefore, the standards are at least somewhat arbitrary. It is by no means clear that the cut-off score used today on the AFQT to determine entrance eligibility is that below which people incapable of satisfactory job performance will score. It would seem, therefore, that the military services must be careful not to place too much emphasis on mental standards.[32] A high score on the AFQT is not necessarily a good indicator of successful job performance.[33]

Although there may be some justification for reducing the number of mental group IV personnel, a tradeoff exists between mental standards and minority accessions. Nearly 40 percent of all Project 100,000 accessions were minorities, mostly black. As we have seen earlier, a large percentage of blacks in the Navy and Marine Corps are in mental group IV, as compared with a much smaller percentage for all enlisted personnel in the re-

---

[31] R. B. Battelle et al., *Analysis of Some Potential Manpower Policies for the All Volunteer Navy* (Menlo Park, Calif.: Stanford Research Institute, 1973), p. 21.

[32] Robert Vineberg and Elain N. Taylor, *Performance in Four Army Jobs by Men at Different Aptitude (AFQT) Levels*, Vol. IV, *Relationships Between Performance Criteria* (Alexandria, Va.: Human Resources Research Organization, 1972). This report concludes that where job performance depends upon a mix of knowledge and skill, then both knowledge tests and performance tests are needed. An instrument, such as a rating scale, that has been designed to assess job motivation and other personal qualities is always essential if a complete estimate of job performance is desired.

[33] As part of a statistical analysis conducted to evaluate majority vs. minority personnel advancement opportunities, the simple correlation (r) was observed between AFQT and job performance for personnel in paygrades E-5 and above. The correlation between job performance and AFQT was found to be weak ($r = 0.06$). More detail is provided in Appendix F.

spective services. The question becomes one of how important minority representation goals are in comparison with other goals such as achievement of a certain level of mental standards among personnel.

OFFICER RECRUITING

The problems associated with recruiting minority officers are different from those associated with recruiting the enlisted. Entrance requirements tend to be less complex, while entry routes tend to be more diverse. Additionally, recruiting methods for attracting officer candidates are different from the methods used in recruiting enlisted personnel. Relatively little officer recruiting takes place in Navy or Marine facilities. Most person-to-person contacts for officer recruiting are made on college campuses by officer recruiting teams.

Educational Attainment

Mental testing is less of an issue in officer recruiting than it is in enlisted recruiting. On the other hand, educational attainment, namely, a four-year college degree, is of paramount importance in becoming an officer. Some officers "come up through the ranks," but the overwhelming majority of officers enter active service at the officer level after graduation from college.

The college degree requirement severely limits the size of the minority manpower pool from which officers may be recruited. According to 1970 census figures, minorities constituted over 16.0 percent of the male population over sixteen years of age, but only 5.9 percent of the college graduates in that age/sex group. Blacks constituted 10.8 percent of the age/sex group, but only 2.9 percent of the college graduates from that group.[34]

College attendance has, however, been on the rise for black males eighteen to twenty-four years of age. In 1971, 20 percent of this age group was enrolled in college, as compared with 11 percent in 1965. The comparable figure for white males remained stable at 34 percent over this period.[35]

Minorities available for the professions continue to be rare. Table III-6 shows that, in 1970, only 2.3 percent of accountants,

---

[34] Data sources are found in U.S., Department of Commerce, Bureau of the Census, *Current Population Reports*, Series P-23, No. 42, "The Social and Economic Status of the Black Population in the United States, 1971," pp. 11-19.

[35] *Ibid.*

TABLE III-6
Employed Persons in Selected Professional Occupations
by Race, 1970

Professional Occupation	Total	Black	Percentage Black	Persons of Spanish Heritage	Percentage Persons of Spanish Heritage
Accountants	703,546	16,521	2.3	15,158	2.2
Engineers	1,207,509	13,679	1.1	25,330	2.1
Lawyers	260,152	3,379	1.3	3,775	1.5
Physicians	280,929	6,106	2.2	10,293	3.7

Source: U.S., Department of Commerce, Bureau of the Census, *Census of Population: 1970, Detailed Characteristics*, Final Report PC(1)-D1, United States Summary, Table 223.

1.1 percent of engineers, 1.3 percent of lawyers, and 2.2 percent of physicians were black. The proportion of persons of Spanish heritage in these professions was only slightly higher in most cases.

Despite much effort to increase the proportion of minorities in various professions, progress is certain to be slow because of the small base—that is, the few minorities who have been historically educated for, or worked in, these professions. Thus, current estimates indicate that, by 1985, blacks will constitute less than 2.0 percent of engineers, 4.8 percent of accountants, 5.7 percent of "other business and management" personnel, 4.5 percent of lawyers, 2.9 percent of physicists, 2.8 percent of chemists, 4.6 percent of physicians, and 4.8 percent of dentists.[36] Even if the most optimistic predictions are made for increased minority entrance into professional schools, the net effect on availability is relatively small. Thus, the gain in black representation in the total engineering labor force as a result of a projected increase in new black entrants from 0.7 percent to 3.5 percent from 1970 to 1985 will be from less than 1.0 percent in 1970 to only 1.8 percent in 1985. Under the best of circumstances, the Navy and

---

[36] See Stephen A. Schneider, *The Availability of Minorities and Women for Professional and Managerial Positions, 1970 to 1985*, Manpower and Human Resources Studies, No. 7 (Philadelphia: Industrial Research Unit, The Wharton School, University of Pennsylvania, 1977), pp. 42, 90, 116, 151, 176, 198, 234, 257.

Recruitment

Marine Corps are certain to have a difficult time recruiting minority personnel for such specialties.

The educational disparity between race-ethnic groups has been compensated somewhat by selection into the officer corps of minorities without college degrees. In fiscal 1974 in the Marine Corps, 75.5 percent of white officers and 67.6 percent of minority officers were college graduates. The same figures for the Navy show 85.0 percent of the white officers and 80.3 percent of the black officers had college degrees. It is interesting to note that, in the Air Force during the same period, 92.3 percent of minority officers were college graduates, compared with 90.6 percent of white officers.[37]

Entry Routes

There are more than fifty programs through which prospective officers may enter the Navy (slightly fewer for the Marine Corps). Most officers, however, come into the services through the programs associated with the following entry routes: Naval Academy, Naval Reserve Officer Training Corps (NROTC), Officer Candidate School (OCS), and Aviation Officer Candidate School (AOCS). Figure III-4 states the age and educational entrance requirements, benefits, and obligations of these programs.

Over 50 percent of the Navy's black officers and more than 40 percent of the black officers in the Marine Corps entered the military through OCS programs. The Army and the Air Force, on the other hand, depend heavily upon ROTC programs in developing minority officer accessions.[38] Tables III-7 and III-8 show minority officer accessions by source of procurement for the Navy and Marine Corps, respectively.

One key difference between white and minority officer procurement patterns lies in the small percentage of minority officers entering the Navy and Marine Corps through the Naval Academy.[39] This difference was most severe in the Navy; in 1974, 18 percent of its white officers, compared with only 3.9 percent of its black officers, were Naval Academy graduates.[40]

---

[37] Scarborough, "Minority Participation," p. 21.

[38] *Ibid.*, p. 20.

[39] Up to 12 percent of a Naval Academy graduating class may select the Marine Corps as a service option.

[40] Scarborough, "Minority Participation," p. 20.

FIGURE III-4
*Overview of Navy and Marine Corps
Officer Entry Routes*

1. U.S. NAVAL ACADEMY; (Men/Women) Line; USN; USMC; Active duty.

AGE—Over 17, but not past 22nd birthday on 1 July of year of entry.

EDUCATION—Normally high school graduate.

SERVICE REQUIREMENTS—Serve on active duty for not less than 5 years upon receipt of original commission.

MARITAL STATUS—Must be unmarried and have no children and remain single upon commissioning.

TRAINING—4 years at the U.S. Naval Academy including summer cruises.

RANK—Midshipman while at the Academy; ENS/USN or 2nd LT/USMC upon graduation.

REMARKS—All candidates must hold nominations. Nominations may be made by the president, vice-president, and members of Congress. In addition, enlisted personnel; children of regular and reserve service personnel (on active duty and been on active duty continuously for 8 years); children of service personnel who died of, or have a service connected disability rated at 100% resulting from, injuries or diseases contracted in active service; and children of personnel who have been awarded Congressional Medal of Honor may receive special consideration for presidential or secretary of the navy appointments. See current Naval Academy catalogue.

2. NROTC MIDSHIPMAN, USNR (4-YR. SCHOLARSHIP PROGRAM); (Men/Women) Line, Supply, Civil Engineer Corps; USN; USMC; Active duty.

AGE—17 on 1 September, but under 21 on 30 June of year of entry. (Not have reached 20th birthday on 30 June if entering five-year program.)

EDUCATION—High school graduate.

MARITAL STATUS—No restrictions.

SPECIAL REQUIREMENTS—Enlist in the Naval Reserve for a period of 6 years; complete such naval science courses, drills, and summer training as may be prescribed; to accept an appointment as a commissioned officer in U. S. Navy or Marine Corps, if tendered. May be released and separated from the program during the first 2 years without prejudice.

SERVICE REQUIREMENTS—Minimum service of not less than 4 years upon receipt of original commission and remain a member of a regular or reserve component of the U.S. Navy or Marine Corps until the 6th anniversary of receipt of original commission.

EDUCATION—Four-year subsidized education at a college or university where NROTC units are established (including at sea/summer training periods).

RANK—Midshipman USNR; ENS/USN or 2nd LT/USMC upon graduating.

Recruitment

FIGURE III-4—Continued

APPLICATIONS—Information available at Navy recruiting stations, high schools, colleges, universities or from Commander, Navy Recruiting Command.

REMARKS—Cost of tuition, associated fees, and books paid by U.S. Navy.

3. NROTC MIDSHIPMAN (COLLEGE PROGRAM, NONSUBSIDIZED); (Men/Women) Line, Supply Corps, Civil Engineer Corps; USNR; Marine Corps Reserve; Active duty.

AGE—17 on 1 September, but under 21 on 30 June of year of entry. (Not have reached 20th birthday on 30 June if entering a five-year college program.) May be 16 if recommended by CO, NROTC Unit.

EDUCATION—High school graduate.

MARITAL STATUS—No restrictions.

SPECIAL REQUIREMENTS—Complete naval science courses, drills and summer training as may be prescribed, and accept a commission as a reserve officer in the U.S. Navy or Marine Corps, if tendered. May be released and separated from the program during the first 2 years without prejudice. Prior to entering the last 2 years, students must enlist in the U.S. Naval Reserve for 6 years.

SERVICE REQUIREMENTS—Serve on active duty for not less than 3 years and serve 3 years inactive duty in Ready or Standby Reserve status.

EDUCATION—4 years college and naval science courses as prescribed at a college or university where NROTC units are established.

RANK—ENS/USNR or 2nd LT/USMCR.

APPLICATIONS—Information available at Navy recruiting stations, high schools, colleges, or universities.

REMARKS—NROTC Unit is responsible for recruiting for this program.

4. OFFICER CANDIDATE SCHOOL (OCS); (Men/Women) Line; USNR; Active duty.

AGE—19 to 27½ at the time of commissioning (adjustable up to 36 months for previous active duty).

EDUCATION—Baccalaureate degree from any regionally accredited college or university.

SERVICE REQUIREMENTS—4 years active duty after commissioning. For men, 6 years Reserve obligation including active and inactive duty.

MARITAL STATUS—No restrictions.

TRAINING—19 weeks at Naval Officer Training Center, Newport, R.I.

RATE/RANK—E-5 (Petty Officer Second Class) until completion of OCS, then ENS/USNR.

APPLICATIONS—May be submitted after completion of junior year of college and when within one year of graduation.

Source: Navy Recruiting Command, *Navy Recruiting Manual—Officer*, COMNAVCRUITCOM Instruction 1110.1A (Arlington, Va.: Department of the Navy, 1975), Chapters III, IV.

TABLE III-7

Minority Officer Accessions by Source of Procurement (Navy)

Source of Procurement	FY 1970 No. (%)	FY 1971 No. (%)	FY 1972 No. (%)	FY 1973 No. (%)	FY 1974 No. (%)
Naval Academy	8 (1.0)[a]	4 (0.5)	9 (1.1)	9 (1.2)	6 (0.7)
ROTC Nonscholarship	13 (1.0)	11 (1.5)	9 (1.5)	1 (0.6)	6 (2.1)
ROTC Scholarship	4 (0.4)	2 (0.2)	11 (1.2)	17 (2.2)	19 (2.0)
Other College Programs	6 (0.7)	4 (0.6)	6 (0.9)	28 (4.3)	4 (1.0)
Reservists	2 (0.4)	2 (0.4)	1 (0.3)	1 (0.4)	1 (0.5)
OCS from Civilian Life	58 (1.3)	94 (3.3)	119 (4.0)	140 (7.7)	75 (5.3)
OCS from College Programs	4 (1.9)	4 (1.6)	2 (0.6)	4 (1.5)	6 (2.4)
Temporary Officer Direct from Ranks	2 (3.6)	—	—	—	—
Other	2 (6.4)	—	1 (2.2)	—	—
Physicians	31 (1.7)	20 (1.4)	19 (1.0)	18 (1.6)	8 (3.5)
Other Medical Specialists	4 (2.8)	7 (3.3)	2 (0.7)	13 (4.7)	6 (2.7)
Senior Medical Students	4 (1.6)	2 (0.7)	1 (0.5)	10 (8.1)	2 (5.7)
Other Direct from Civilian Life	2 (0.9)	—	4 (1.1)	4 (0.9)	8 (3.3)
Total Commissioned Officer	140 (1.2)	150 (1.7)	184 (2.0)	245 (3.7)	141 (2.8)

Source: Edward Scarborough, "Minority Participation in the DOD" (Washington, D.C.: Defense Manpower Commission, 1976), Appendix A.

[a] Percentage of those officers procured through this source. For example, in fiscal 1970, 1.0 percent of Navy officers procured through the Naval Academy were minorities.

Recruitment

TABLE III-8
Minority Officer Accessions by Source of Procurement (USMC)

Source of Procurement	FY 1970 No. (%)	FY 1971 No. (%)	FY 1972 No. (%)	FY 1973 No. (%)	FY 1974 No. (%)
Naval Academy	3 (2.8)[a]	1 (0.8)	8 (3.3)	4 (2.8)	3 (3.3)
ROTC Nonscholarship	2 (3.4)	2 (6.9)	2 (5.9)	3 (14.3)	4 (16.7)
ROTC Scholarship	2 (2.1)	3 (3.5)	3 (3.1)	4 (3.3)	10 (5.9)
Other College Programs	9 (1.1)	8 (1.1)	21 (4.1)	14 (2.3)	27 (4.5)
Reservists	1 (1.0)	1 (0.9)	1 (2.7)	—	—
OCS from Civilian Life	27 (2.1)	14 (1.8)	36 (4.7)	49 (5.8)	51 (8.0)
OCS from Active Military	10 (2.7)	1 (1.5)	2 (2.6)	5 (6.6)	1 (1.7)
OCS from College Programs	—	—	2 (15.4)	—	1 (4.0)
Temporary Officer Direct from Ranks	15 (11.4)	2 (8.0)	—	21 (10.0)	11 (10.7)
Total Commissioned Officers	69 (2.3)	32 (1.7)	75 (4.4)	100 (4.9)	108 (6.1)

Source: Edward Scarborough, "Minority Participation in the DOD" (Washington, D.C.: Defense Manpower Commission, 1976), Appendix A.
[a] Percentage of those officers procured through this source. For example, in fiscal 1970, 2.8 percent of Marine Corps officers procured through the Naval Academy were minorities.

Recalling that the vast majority of officers entering the services have college degrees and that most minorities enter the Navy and Marine Corps directly from civilian life through OCS, it is clear that these services must compete for minority college graduates. As noted earlier, there is no surplus of minority college graduates. Additionally, private industry, the professions, and the various levels of government actively compete for the services of this small manpower pool. It is doubtful that the Navy and Marine Corps will be able to increase minority officer accessions significantly in the near future if they continue to concentrate so much effort on minority degree holders.

An alternative to the above method of officer recruiting would be to increase minority officer accessions through NROTC and the Naval Academy. This would shift the recruiting effort from the minority college graduate to minorities in high school or the first year or two of college. This approach allows the military to recruit a prospective minority college graduate before his services are in such great demand.

The Naval Personnel and Training Laboratory (now called Naval Personnel Research and Development Center) conducted a study to evaluate the results of a minority officer recruitment effort which emphasized early detection of potential officer recruits.[41] The Strong Vocational Interest Blank was administered to 460 high school juniors in late 1970. Ninety-six of these high school students, including 33 minorities, scored relatively high on the Navy interest scales. Interviews were arranged between these students and a Naval Academy information officer. About 40 of these students, including 12 minority students, indicated to the interviewer a positive interest in a naval career. The interviews revealed that a large number of these students had not previously considered the Navy among their career possibilities.

Interviews were again conducted in late 1971. Ninety-one of the ninety-six students previously interviewed were reinterviewed. Thirteen of these students had applied for either an NROTC scholarship or an appointment to the Naval Academy. Unfor-

---

[41] See Patricia J. Thomas and Bernard Rimland, *The Use of a Vocational Interest Test in Recruiting Minority and Caucasian Officer Candidates: An Exploratory Study*, SRM 72-3 (San Diego, Calif.: Naval Personnel and Training Research Laboratory, 1971).

tunately, four of the five students who could not be reinterviewed were minorities. Of the thirteen individuals who did apply for a scholarship, two were minorities.[42] Certainly, high school students' awareness of the officer recruitment programs available would be increased by such a program. Very likely, this early stage minority penetration approach would increase the number of minorities applying to the NROTC and Naval Academy programs.

---

[42] See Patricia J. Thomas and Don G. Hinsvark, *An Evaluation of a Vocational Interest Test in Recruiting Officers*, SRR 73-16 (San Diego, Calif.: Naval Personnel and Training Research Laboratory, 1973).

CHAPTER IV

A Statistical Analysis of Minorities' Upgrading Opportunities in the Enlisted Navy

Our statistical analysis was conducted to accomplish two objectives: first, to model the Navy enlisted personnel advancement functions based on both personal characteristics and in-service variables;[1] and second, and more important, to evaluate minorities' vs. nonminorities' advancement opportunities with respect to the variables found to be significant. The succeeding chapters describe the systems of occupational classification, assignment, promotion, and retention and analyze their current impact on upgrading minority personnel. Is proportionate representation across paygrades and occupations attainable given the institutional policies now in practice and the educationally disadvantaged status of the minority manpower pool? This question is answered, in part, by the conclusions derived from the statistical analysis. This analysis draws attention to the variables which define the enlisted advancement function and which have an adverse impact on minorities' upgrading opportunities.

Stepwise multiple regressions were used to model the advancement functions. The variables which were considered for these advancement functions are defined in Table IV-1. Frequency distribution and correlation analyses were used to compare minorities' and nonminorities' profiles with respect to the significant variables in the models. Model interpretations based on these regression and correlation analyses follow the descriptions of each model. Information concerning model construction and bias, statistical results, and methodology is not presented here, but rather is discussed in the Appendixes.

---

[1] The variables used in the analysis were those thought to be important and which could be obtained from the Department of Defense.

TABLE IV-1
Definition of Advancement Function Variables

Variable	Description
1. Paygrade	Member's paygrade (rank) as of June 1975. E-1 is the lowest paygrade; E-9 is the highest. Paygrade is the dependent variable in the regression models.
2. Armed Forces Qualification Test (AFQT)	The actual AFQT was discontinued in January 1973; however, the scores derived from current tests to determine enlistment eligibility are converted to, and reported as, an AFQT percentile score.
3. General Classification Test (GCT)	One of six sections of the Basic Test Battery (BTB).[a] The GCT measures verbal comprehension, which entails the ability to understand written and spoken language, thus indirectly measuring reasoning ability. It is represented most heavily in what is often termed as reading skill. Vocabulary is only a factor which characterizes reading skill; but it provides a good measure of verbal comprehension.
4. Arithmetic Reasoning Test (ARI)	Designed to measure general reasoning, it is concerned with the ability to generate solutions to problems. ARI tests the ability to use numbers and apply mathematical reasoning in practical problems.
5. Mechanical Comprehension Test (MECH)	Tests aptitude for mechanical work, mechanical and electrical knowledge, and ability to understand mechanical principles.
6. Clerical Test (CLER)	Tests the ability to observe details rapidly and measures the speed of responses to observations.
7. Shop Practices Test (SHOP)	Measures functional ability of an individual who has had experience with, and is knowledgeable about, the use of a variety of tools found in a shop. The experience with, and knowledge of, shop practices might be acquired from high school shop courses.

TABLE IV-1—Continued

Variable	Description
8. Years of education completed	A member's number of years of education completed for which credit was received.
9. Marital status	Individual designated either married or single, divorced, legally separated, widowed, or marriage annulled.
10. Age at entry	Age at which individual entered the Navy.
11. Black personnel	Individual is member of Negro race.
12. Other minority personnel	Individual is member of one of the following ethnic groups:[b] Caucasian of Spanish descent, American Indian, Asian American, Puerto Rican, Filipino, Mexican American, Eskimo, Cuban American, Chinese, Japanese, Korean.
13. Region of residence	Indicates an individual's official residence at time of enlistment. The United States was divided into eight regions as follows: Central Atlantic, Pacific, Southeast, Great Lakes, Southwest, Midwest, Northeast, and Noncontinental United States.
14. Time in service	Number of months an individual has been in the Navy.
15. Discipline record	Indicates whether or not individual has been disciplined to the extent of a reduction in rate.
16. Occupations with open advancement potential[c]	Includes those ratings which provide the greatest amount of advancement opportunity because of manpower shortages in the petty officer enlisted rates (E-4 and above).
17. Occupations with closed advancement potential	Includes those occupations which provide the least amount of advancement opportunity due to overmanning in paygrades E-4 and above.

Statistical Analysis

TABLE IV-1—Continued

Variable	Description
18. Current evaluation variables[d]	The following variables provide information on a command's evaluation of a member's personal performance and abilities: performance, appearance, cooperativeness, reliability, conduct, resourcefulness, leadership, overall and equal opportunity evaluation (member's ability to deal with individuals of all race-ethnic groups in a nondiscriminatory manner).

[a] The Basic Test Battery (BTB) consists of the following sections: GCT, ARI, MECH, CLER, SHOP, and Electronics Technician Selection Test (ETST). The definitions are taken from Bureau of Naval Personnel, *Manual of Enlisted Classification Procedures*, NAVPERS 15812B (Washington, D.C.: Department of the Navy, 1970).

[b] The other minority group designations are those used by the Department of Defense. See Navy Recruiting Command, *Navy Recruiting Manual—Enlisted*, COMNAVCRUITCOM Instruction 1130.8A (Washington, D.C.: Department of the Navy, 1974), pp. 3-10.

[c] Career Reenlistment Objectives (CREO) is a personnel management system designed to match manpower requirements and advancement opportunity with occupational categories. The three CREO occupational categories are open rating, neutral rating, and closed rating. Personnel not assigned to either open or closed occupations must be, by elimination, assigned to neutral occupations. The neutral rating occupations represent an approximate match between manpower requirements and personnel manning. Navy ratings (occupations) are broken down into CREO occupational categories in Bureau of Naval Personnel, "Career Reenlistment Objectives (CREO)," BUPERS Instruction 1133.25C (Washington, D.C.: Department of the Navy, 1975).

[d] Variables are fully defined in Bureau of Naval Personnel, NAVPERS 1616 Series; and *Bureau of Naval Personnel Manual*, NAVPERS 15791B (Washington, D.C.: Department of the Navy, 1969).

The analysis was conducted in two phases. In the first phase, we used preservice information to model the Navy enlisted advancement function. In the second phase, we combined preservice information with data on in-service characteristics. It is of value to the Navy to look at race-ethnic group comparisons based on information collected at enlistment and shortly thereafter and again when in-service performance and occupational data are available. Both models generated their results from a common sample of individuals whose range of time in the Navy was two months to fifteen years.

The preservice plus in-service model was later refined in several ways. First, separate models were constructed for blacks and for other minorities.[2] The black and other minority samples used for those models were derived from the original sample. The above models were constructed from a single snapshot of the sample personnel rather than from several snapshots over a period of time. An additional refinement involved the construction of a cross-sectional model to control for time in the Navy. All persons included in this sample had approximately four years of active duty as of the file date. The final refinement expanded the original preservice plus in-service model to include performance evaluation variables. This information was available only for those in paygrades E-5 and above. Therefore, a new sample was constructed for this model. To be included in the sample, an individual must have been in the E-5 paygrade or higher. The range of time in the Navy for members of this sample was thirteen months to sixteen years.

PRESERVICE MODEL

An advancement function was constructed from personal and environmental characteristics compiled on personnel at enlistment and shortly thereafter. The variables which make up the preservice advancement function are Armed Forces Qualification Test (AFQT) score,[3] number of years of education completed, all subtests of the Basic Test Battery (BTB),[4] official region of residence at time of enlistment, marital status, race-ethnic characteristics, and age at entry.

One objective of this model was to determine which, if any, of the above variables contribute to the explanation of paygrade

---

[2] Other minority personnel comprises members of the following race-ethnic groups: Caucasian of Spanish descent, American Indian, Asian American, Puerto Rican, Filipino, Mexican American, Eskimo, Cuban American, Chinese, Japanese, and Korean.

[3] The term *AFQT score* is still used by the Navy and Marine Corps although the AFQT (Armed Forces Qualification Test) itself was discontinued in January 1973. The AFQT score is derived from either the Basic Test Battery (BTB) or Armed Services Vocational Aptitude Battery (ASVAB), depending on which test was used to determine enlistment eligibility.

[4] The BTB, administered at boot camp, consists of the following subsections: General Classification Test (GCT), Arithmetic Reasoning Test (ARI), Mechanical Comprehension Test (MECH), Clerical Test (CLER), Shop Practices Test (SHOP), and Electronics Technician Selection Test (ETST). An individual's occupational classification is determined, to a large extent, by his composite score on the BTB. The series of exams in the BTB ostensibly combine to provide accurate information on an individual's mental aptitude and vocational aptitude.

level attainment. Of particular interest was whether or not the race-ethnic variables are significant and, if so, to what extent. That is, when the contributing explanatory variables other than race-ethnic information are held constant, do minorities advance as quickly as nonminorities? The second objective of the model was to compare minorities' and nonminorities' profiles with respect to the variables in the model.

Interpretation of Preservice Model Statistical Results

The following factors, listed in descending order of significance, are found to have a statistically significant relationship with the dependent variable: Arithmetic Reasoning Test (ARI) score, marital status, years of education completed, other minority personnel, General Classification Test (GCT) score, black personnel, age at entry, AFQT score, Shop Practices Test (SHOP) score, Clerical Test (CLER) score, and the Central Atlantic, Southeast, and Great Lakes regions of residence. The first five factors are highly significant, and each shows a strong relationship with paygrade level attainment. Black personnel, age at entry, AFQT, and SHOP show less of a relationship. The remaining variables each have a weak, but statistically significant, relationship with the dependent variable. The model statistics are shown in Table IV-2.

With the exception of other minority personnel and black personnel, the above factors are *positively* related to paygrade attainment.[5] Therefore, in the aggregate, an individual with a higher ARI score will be promoted faster than an individual with a lower ARI score when all other significant factors are identical. The same holds true for an individual with a higher GCT, AFQT, and/or SHOP score, as well as for an individual who enters the Navy at an older age and/or is married.

Generally speaking, the results of this model show that minorities are promoted more slowly than nonminorities.[6] Even when the significant variables other than race-ethnic are held constant, blacks and other minority personnel are not promoted quite as quickly as majority personnel. That is to say, in the aggregate, a minority individual is not promoted at the same rate as a

---

[5] The variables "Southwest region" and "Great Lakes region" are also negatively related; however, these two variables are of lesser significance.

[6] Dummy variables were used to distinguish black and other minority personnel from majority personnel. Negative coefficients on the dummy variables imply slower advancement rates.

TABLE IV-2
Preservice Model Statistics: Significant Variables Listed in Order of Significance

Variable	Standard Regression Coefficient	Contribution to R^2
ARI	0.234	0.3279
Marital status	0.370	0.1492
Years of education	0.189	0.0250
Other minority personnel	—0.107	0.0100
GCT	0.102	0.0091
Black personnel	—0.075	0.0047
Age at entry	0.056	0.0022
AFQT	0.103	0.0021
SHOP	0.063	0.0029
CLER	0.032	0.0008
Central Atlantic region [a]	0.023	0.0006
Southeast region [a]	—0.026	0.0006
Great Lakes region [a]	—0.022	0.0005
		0.5357

[a] Variable is significant at the 0.025 level of significance. All other variables are significant at the 0.0005 level of significance.

majority individual even though both are from the same region, enter the Navy at the same time and age, have the same number of years of education completed, and receive equivalent scores on the BTB and the AFQT. In addition, blacks are promoted slightly faster than other minority personnel.

The above preservice significant variables account for much of the race-ethnic groups' promotion rate differences. Those variables which are also inextricably tied to a policy framework in occupational classification, assignment, promotion, and/or retention are of particular interest. It is important to determine to what extent group promotion differences are explained by these variables.

Academic credentials, as a whole, are found to be extremely important factors with respect to advancement opportunity. ARI and GCT scores [7] and the number of years of education completed are the most important of the academic variables. It can be safely said that individuals possessing capabilities which are measured by higher scores on the ARI and GCT have a distinct advantage in advancement opportunity. Of the Basic Test Battery, CLER, Mechanical Comprehension Test (MECH), and SHOP are the least influential on the advancement function. MECH is not significant, and CLER and SHOP are only weakly significant.

The dominant factors determining Navy enlistment eligibility are mental aptitude (measured by either a short version of the BTB or the AFQT) and educational level. The composite score obtained on the BTB is the dominant factor in determining occupational classification. It is not surprising, then, that ARI, GCT, and AFQT scores, as well as number of years of education completed, are significant variables in the model.[8]

Only small differences in number of years of education completed are found between race-ethnic groups.[9] Therefore, the years of education variable accounts for very little of the difference in the groups' promotion rates. On the other hand, there are large differences in the mental aptitude exam scores between majority and minority personnel. In the aggregate, minorities, especially blacks, have much lower AFQT, GCT, and ARI scores than nonminorities.[10] These variables are extremely important for promotion opportunities. The fact that these variables are heavily weighted, coupled with the large differences in race-ethnic group mean scores, explains to a large degree the slower promotion of minority personnel.

[7] The ARI and GCT are the most heavily weighted sections of the BTB for occupational classification.

[8] The correlation between paygrade level and the highly significant academic variables are as follows: GCT, .5226; ARI, .5742; AFQT, .5228; and years of education completed, .4562. It is clear from these correlations that academic credentials are extremely important for advancement.

[9] The mean years of education completed for the aggregate sample, blacks, and other minority personnel, respectively, are 11.8, 11.5, and 11.2 years. The proportion of variation of the years of education variable explained by the black personnel variable is 0.6 percent. The proportion of variation of the years of education variable explained by other minority personnel is 2.2 percent.

[10] The aggregate sample mean scores for AFQT, GCT, and ARI, in that order, are 64.0 (s.d. 21.0), 56.0 (s.d. 9.7), and 53.6 (s.d. 9.0). The mean scores for blacks on these same examinations are 43.0, 46.0, and 44.7.

The above results might lead one to question the educational *quality* differences between minority and majority personnel. The model results indicate a strong positive relationship between years of education and the mental aptitude exams.[11] Furthermore, there is additional evidence to support the contention that educational level is highly correlated with so-called mental aptitude.[12] Given that the years of education of minority personnel relative to majority personnel are approximately equal, perhaps the quality of education of minority personnel, again relative to majority personnel, accounts for the group differences in aptitude scores, as well as a great deal of the variation with respect to promotion rates.

If differences in quality of education account for much of race-ethnic group promotion rate disparities, then the Navy can choose one of two alternatives in order to upgrade minority personnel. First, the service can downgrade the relative importance of academic credentials in the advancement function. This is one way to decrease race-ethnic groups' promotion rate differences. (Downgrading the weight of academic variables as a means to achieve proportionate representation of minority personnel has already been touched on in the last chapter; the viability of such a decision is further discussed in chapters V and VI.) Second, the Navy can provide a remedial education program for those who have potential, but who are educationally disadvantaged. Such a program would certainly have a positive effect on minorities' advancement opportunities. (Remedial education as a means to upgrade minority personnel is examined in the next chapter.)

The preservice advancement model is a satisfactory one. The combination of statistically significant independent variables explains a great deal of the variation in paygrade level attainment.[13] Furthermore, the model provides insight into which personal and environmental characteristics contribute, and to what extent, to the explanation of paygrade level attainment.

---

[11] The correlations between years of education and AFQT, GCT, and ARI, respectively, are .42, .50, and .49.

[12] See Arthur I. Siegel, Brian A. Bergman, and Joseph Lambert, *Nonverbal and Culture Fair Performance Prediction Procedures*, Vol. II, *Initial Validation* (Wayne, Pa.: Applied Psychological Services, Inc., 1973), pp. 2-3.

[13] The multiple correlation (R^2) measures the proportion of the total variation of the dependent variable which is explained by the regression equation. For the preservice model, R^2 is .5357.

Statistical Analysis

The model indicates that minorities are not promoted as quickly as nonminorities. When the preservice significant variables other than race-ethnic are held fixed, however, much, but not all, of the race-ethnic groups' promotion rate differences are accounted for. The remaining differences in minority vs. majority and in black vs. other minority promotion rates (i.e., differences in advancement after preservice variables are controlled) must be accounted for by one or both of two possible reasons. First, the promotion rate differences might be accounted for by other personal and environmental characteristics which are not compiled on an individual at enlistment and, therefore, are not considered in this model. Examples of additional factors which, if measured, might provide more insight into an individual's promotion potential are motivation and vocational interest. These factors might also help to explain group promotion differences further. Finally, the unexplained promotion rate differences might be the result of in-service phenomena.

PRESERVICE PLUS IN-SERVICE MODEL

One reason for constructing the preservice plus in-service model was to determine whether additional information such as occupational category, discipline record, time in the Navy, and on-the-job evaluations add, and to what extent, to the explanation of paygrade level attainment. Do the significant preservice variables of the initial model remain significant with the addition of in-service variables? It will be interesting to determine the mix of significant preservice and in-service variables which constitute the advancement opportunity model. Do the race-ethnic variables remain significant after the in-service variables are introduced and controlled for? The second reason for enlarging the preservice model was to compare minority and majority personnel profiles with respect to the significant in-service variables in the model. What is the relative impact of the above variables on minorities' advancement?

The independent variables that were considered for the preservice model remained as candidates in this model. In addition, occupational classification, discipline record, and time in the Navy were considered. As before, paygrades E-1 through E-7 are represented in the sample.

Neither advancement exams nor performance evaluations were considered in this model. Advancement examination scores could not be obtained, and performance evaluations were available only for paygrades E-5 and above. Performance evaluation scores are considered in a later model in which the sample consists of personnel in paygrades E-5 and above.

Interpretation of the Preservice Plus In-Service Model

The following factors are found to be significant in decreasing order of significance: time in the Navy, ARI, discipline record, GCT, years of education, AFQT, occupations with open advancement potential,[14] CLER, occupations with closed advancement potential, marital status, other minority personnel, and SHOP. The first five variables are highly significant and contribute substantially to the explanation of the dependent variable's variance. They account for 97 percent of the multiple correlation of the model. The remaining variables are statistically significant, but display a relatively much weaker relationship with the dependent variable. All factors are positively related to paygrade level with the exception of discipline record, occupations with closed advancement potential, and other minority personnel. These factors are negatively related to the dependent variable. The model statistics are shown in Table IV-3.

The following preservice variables continue to demonstrate a statistically significant relationship with the dependent variable despite the incorporation of in-service variables: ARI, GCT, years of education, AFQT, CLER, marital status, other minority personnel, and SHOP. These variables continue to affect an individual during the course of his enlistment. Note that the majority of academic variables remain significant.

The black personnel variable is no longer significant. This means that blacks are promoted as quickly as members of the

---

[14] Career Reenlistment Objectives (CREO) is a personnel management system designed to match manpower requirements and advancement opportunity with occupational categories. The three CREO occupational categories are open rating, neutral rating, and closed rating. The open rating category (fast promotion rate occupations) includes those occupations which provide the greatest amount of advancement opportunity because of manpower shortages in the petty officer enlisted rates (E-4 and above). The closed rating category (slow promotion rate occupations) includes those occupations which provide the least amount of advancement opportunity due to overmanning in paygrades E-4 and above. Bureau of Naval Personnel, "Career Reenlistment Objectives (CREO)," BUPERS Instruction 1133.25C (Washington, D.C.: Department of the Navy, 1975).

Statistical Analysis 83

TABLE IV-3
Preservice Plus In-Service Model Statistics: Significant Variables Listed in Order of Significance

Variable	Standard Regression Coefficient	Contribution to R^2
Time in Navy	0.666	0.6853
ARI	0.104	0.0718
Discipline record	—0.139	0.0208
GCT	0.072	0.0101
Years of education	0.084	0.0055
AFQT	0.075	0.0036
Open occupations	0.040	0.0026
CLER	0.044	0.0016
Closed occupations	—0.040	0.0014
Marital status	0.043	0.0013
Other minority personnel	—0.036	0.0012
SHOP	0.031	0.0007
		.8059

Note: All variables are significant at the 0.005 level of significance.

majority when the significant preservice plus in-service variables are held fixed. So, for a given black and a given majority member with *identical* academic credentials, occupational classifications, discipline records, and time in the Navy, there will be no difference in promotion rates between the two individuals. This does not mean that black personnel, in general, are promoted as quickly as majority personnel. In fact, the preservice model results clearly state otherwise. Yet, as stated, the results of this model show that a black individual and a majority individual with *identical* preservice and in-service variable characteristics will advance at the same rate. Since the black and majority personnel populations *are not identical*, the important question becomes, Which of the statistically significant variables accounts for the promotion rate disparities? This question will be examined shortly.

The other minority personnel variable remains significant in the preservice plus in-service model. Thus, other minorities are promoted slightly more slowly than majority personnel (and blacks) when the significant variables are held fixed. That is, in the aggregate, given an individual classified as an other minority and an individual classified as a nonminority who have equivalent time in the Navy, academic credentials, marital status, discipline records, and occupational classifications, the majority member will be promoted faster than the other minority member. The other minority personnel factor has, however, a much weaker relationship with paygrade level when the in-service variables are included in the model. It will be interesting to see whether incorporating the current evaluation data into the E-5 and above model will account for the remaining difference in promotion rates.

In the preservice model, black personnel and other minority personnel were found not to have achieved as high a paygrade level as majority personnel. With the addition of the in-service variables into the model, however, black personnel is no longer a significant variable, and the difference between other minorities' and nonminorities' promotion rates when the significant variables are controlled for has been reduced. Therefore, when variables which measure in-service phenomena are included in the model and when the differential promotion rate evaluations are made based on identical race-ethnic group scores for these variables, blacks are found to be promoted at the same rate as majority personnel, and other minorities are found to be promoted at a slightly slower rate.

The combination of in-service and preservice variables completely explains the advancement rate differences between black and majority personnel and goes a long way towards explaining those differences between other minority personnel and majority personnel. All of the in-service variables which were entered into the model are statistically significant. The question still remains, however, Which of the statistically significant variables accounts for the promotion rate disparities?

In the aggregate, an individual with more time in the Navy will have a higher paygrade than an individual with less. The range of time in the Navy for members of the sample used to generate both this model and the preservice model was two months to fifteen years. The frequency distribution of time in service by race-ethnic group provides interesting results. Twenty-eight percent of the aggregate sample, or somewhat less than that for majority personnel, had lengths of service as of the file date of less than or equal to twelve months. Approximately 50 percent of the minorities in the sample, however, had equivalent lengths of service. Generally speaking, minorities in the sample had less time in the Navy than nonminorities had. As already stated, time in the Navy has a very strong positive relationship with paygrade level attainment. Certainly the differences in the race-ethnic groups' average lengths of naval service account for much of the difference in their average paygrade levels.

From the model, it is evident that the decision whether or not an individual is advanced depends, to a large extent, on his disciplinary record. The percentage of the sample reduced in paygrade because of disciplinary actions is 2.5. The percentages of black personnel and other minority personnel reduced in paygrade are 5.3 and 3.3, respectively. Thus, the percentages of blacks and other minority personnel reduced in paygrade exceed the aggregate sample percentage. Because disciplinary action is a factor which is strongly related to paygrade level, the impact on minority advancement of more than proportionate "busts" is severe. (The military justice system is investigated in chapter VI to determine, first, whether justice is distributed fairly and, second, whether all groups *perceive* the system as fair.)

Open and closed advancement opportunity occupations are statistically significant variables. Occupational classification, therefore, does make a difference in an individual's promotion rate. Individuals assigned to the so-called open occupational ratings are promoted slightly faster than individuals assigned to the neutral occupational ratings, all else being equal. In turn, those in neutral ratings are promoted slightly faster than those in closed ratings. In the sample, 42.0 percent of the personnel are assigned to an open rating, 50.0 percent to a neutral rating, and 8.0 percent to a closed rating. Concerning black personnel in the sample, approximately 27.0 percent are assigned to open ratings, 70.0 percent to neutral ratings, and approximately 2.0 to 3.0 percent to closed ratings. With respect to the other minority personnel, 41.8 percent are assigned to open ratings, 51.6 percent to neutral ratings, and 6.6 percent to closed ratings.

It is difficult to evaluate the overall impact on minority personnel of the above distribution of occupational classification. Blacks are severely underrepresented in the occupations with the greatest advancement potential. Blacks are also underrepresented in the slowest advancement categories, but overrepresented in the neutral advancement track. The net conclusion is that blacks are not proportionately distributed across occupations nor as well off in advancement opportunity relative to majority personnel as a result of this distribution. Other minorities are distributed approximately proportionately across advancement opportunity job tracks.

The model is a good one. The multiple correlation increases substantially as a result of entering the in-service variables into the model.[15] Academic credentials, occupational classification, time in service, and discipline record are jointly found to be strongly related to the enlisted advancement function. Correlation and frequency distribution analyses clearly indicate that minorities are adversely affected, in relative terms, by the strong influence of these variables on paygrade level attainment.

SEPARATE BLACK AND ETHNIC MODELS

Separate preservice plus in-service models for both black personnel and for other minority personnel were derived from the original sample in order to compare the race-ethnic models with each other and each with the aggregate model. The black subsample consists of the 524 blacks in the sample, and the ethnic subsample consists of the 390 other minorities in the sample.[16] The explanatory variables that were considered for the original preservice plus in-service model remained as candidates in these models with the exception, of course, of the race-ethnic variables.

Comparison of Original, Black, and Other Minority Personnel Models

The variables found in the original preservice plus in-service model to influence promotion rates essentially remain the same when separate models are run for black and other minority personnel. There were some exceptions, however. Furthermore,

---

[15] For the preservice plus in-service model, R^2 is .8059.

[16] From the Bureau of Naval Personnel master enlisted file tape, a random sample of 5,000 cases (individuals) was extracted. Minorities are well represented in this sample (i.e., 10.6 percent black personnel and 7.8 percent other minority personnel).

although many of the influential variables are the same, the degree of variable significance for the separate models is less than that of the original model.[17]

In the black model, the following variables are found to be significant in decreasing order of significance:[18] time in the Navy, discipline record, AFQT, years of education completed, SHOP, marital status, GCT, and occupations with open advancement potential. In the other minority model, the following variables, also listed in decreasing order of significance, are statistically related to paygrade: time in the Navy, AFQT, discipline record, years of education, occupations with closed advancement potential, and ARI. The first three of these variables account for 95 percent of model multiple correlation. Within the black personnel model, the first four variables strongly influence promotion and together account for 97 percent of model correlation. The remaining variables each account for a relatively weak, but statistically significant, relationship with paygrade level.

Concerning the in-service variables, the original model results indicate that time in the service and discipline record are most strongly related to paygrade. The occupational classification variables were each found to display a much weaker relationship with the dependent variable. In the separate race-ethnic models, time in the service and discipline record remain the most influential in-service variables; however, whereas both occupational classification variables were found to be significant in the original model, only open occupation in the black model and only closed occupation in the other minority model remain significant.

With respect to the preservice variables, recall that the results of the original model indicated that the joint significance of the BTB, AFQT, and years of education completed is great. In that model, these academic variables contribute, in large part, to the explanation of paygrade level variation; and of the academic variables, ARI and GCT display the strongest relationship with the dependent variable.

---

[17] R^2 for the black regression equation is .7470, and for the final other minority personnel equation, .5289. R^2 for the original model is .8059. An explanation of the differences in the value of multiple correlation between the black, other minority, and original models is presented in Appendix D.

[18] The black and other minority models statistics are found in Appendix D (Tables D-1 and D-2, respectively).

When separate models are run for black personnel and for other minority personnel of the sample, the joint significance of the academic variables is not found to be as important. In both models, the AFQT shows the strongest relationship with paygrade level. In the black model, GCT is the only other statistically significant exam variable, and in the other minority model, ARI is the only other significant exam variable. Both GCT, for blacks, and ARI, for other minority personnel, display relatively weak relationships with the dependent variable.

These results of the two race-ethnic models contrast with those of the original model, in which GCT and ARI are strongly influential and AFQT less so. A possible explanation lies in the services' practice of guaranteeing occupational classifications to a certain percentage of new recruits at enlistment. Until recently, the selection criterion for this guaranteed formal school assignment was the AFQT. Normally, however, selection for formal school training was determined by the BTB exams, especially the GCT and ARI subtests. Perhaps minority recruits were more than proportionately represented in the pool of those provided with guaranteed formal school assignments. This is quite possible as a result of the recent emphasis on recruiting qualified minority personnel. If so, then AFQT, the guaranteed assignment selection criterion, became more important for minority personnel than for majority personnel in determining occupational classification. Occupational classification does influence promotion rate; therefore, for minority personnel especially, AFQT was an influential variable in determining paygrade.

CROSS-SECTIONAL MODEL

A cross-sectional study was conducted to supplement the original preservice plus in-service model by controlling for the number of months in the Navy. By controlling for this variable, the strong influence of time in the Navy was neutralized. Therefore, one objective of this cross-sectional model was to determine the mix of significant variables when time in the Navy is held constant and to answer the questions, Do the original and cross-sectional models provide the same results on minorities' vs. nonminorities' advancement? Individuals selected for the cross-sectional sample were screened in the same manner, with one exception, as those individuals included in the original preservice plus in-service sample. The exception was that the individuals included in the cross-sectional sample entered the Navy in either April, May, June, or July 1971. Therefore, this model constructs an advancement function for individuals who had approximately four years of active naval service as of the file date.[19]

The variables studied in this model were the same as those studied in the preservice plus in-service model. Time in the Navy was still considered in order to accommodate the difference of up to three months which was possible. The relative significance of this variable was expected, however, to be much less than found in previous models. The multiple correlation, therefore, was expected to drop considerably.

Cross-Sectional Model Interpretation

Everyone in the cross-sectional model served in the Navy as an enlisted man between 1971 and 1975 as of the file date. Therefore, the results of the cross-sectional model are based on the situation existing during this period. The range of time in the Navy for members of the sample in the original model was two months to fifteen years, although 50 percent of the individuals in the original model had no more than twenty-one months of active service. Because of the frequency distribution of time in the service, the original model describes the advancement function most reliably from 1973 to 1975. It is clear that the range and median values for time in the Navy are very different for the original and cross-sectional samples. Because the time frames are different, the sample paygrade distributions are different. Thus, because of the disparate samples from which the cross-sectional and original models were derived, the models' results were not expected to be the same. Model comparisons, then, provide for a further understanding of the Navy enlisted promotion function, as well as for a further understanding of the relative impact of the significant variables on race-ethnic groups.

Recall that in the original preservice plus in-service model, other minority personnel was found to be a significant variable, but the black personnel variable was not. When the significant variables were controlled for, other minorities were found to be promoted slightly more slowly, in the aggregate, than either

---

[19] Four years of active duty is the normal length of time for an enlistment contract.

black or majority personnel. Black and majority personnel were found to be promoted at the same rate.

In the cross-sectional model, however, the sample consists of individuals who have four years of active service. Interpretations are based on the results of a specific four-year period. Other minority personnel is no longer a significant variable. The black personnel variable is statistically significant, but has a positive relationship with paygrade which is weak relative to the other significant variables in the equation. Nevertheless, when the significant variables are held fixed, those blacks still in the service after four years of obligated service are found to be promoted at a slightly faster rate than majority personnel and other minority personnel. That is, in the aggregate, given a minority member and a majority member who entered the Navy in 1971 and completed their obligated service and have identical academic credentials, marital status, time in paygrade, discipline records, and occupational classifications, the minority member will be promoted at least as quickly as the majority member. This phenomenon reflects the emphasis the Navy has placed on minority upgrading since 1972.

The average academic credentials scores of personnel in the cross-sectional sample are found to be higher than those of personnel in the original sample. The cross-sectional sample has a greater value for mean time in the Navy than the original sample. Thus, the sample with the greater value for mean time in the Navy has a higher average value for academic credentials scores than has the sample with a lower value for mean time in the Navy. During the four-year period considered by the cross-sectional model, a great deal of attrition took place. The armed services reduced their manpower requirements considerably in an effort to scale down after Vietnam. The results of the model indicate that the individuals who left the service generally had less impressive academic credentials than had those who remained.

The period between 1971 and 1975 was one of considerable flux in minority-upgrading policy and manpower requirements in the Navy. The percentage of minority personnel recruited fluctuated considerably during this period. The Navy established minority personnel recruiting and upgrading goals, including occupational classification priorities for qualified minorities. The scaling down of manpower requirements also took place. As a result, individuals were discharged prior to completion of obligated service. The attrition rate of those individuals having lower academic credentials was higher than that of those having higher academic credentials. Results indicate that the attrition rate for minority personnel was higher than for majority personnel. It is interesting to note, however, that, when the statistically significant cross-sectional variables are held fixed, minorities who remained in service, in the aggregate, are found to be promoted slightly faster than majority personnel.

Generally, interpretations based on the original preservice plus in-service model results are valid here. That is, for the most part, variables found to contribute to the description of the original advancement model also explain the cross-sectional advancement function. Additional insight, however, is provided through model comparisons where the sample time frames and paygrade segments are different.

Although for the most part the variables found to be statistically significant in the original model remain significant in this model, the joint significance of these variables is less than that in the original model.[20] As already mentioned, the multiple correlation was expected to drop considerably as a result, to a large extent, of controlling time in the Navy. The relative lack of significance is also partly due to the difference in personnel-upgrading policy during these time periods. Furthermore, the academic credentials differences between individuals are not as great in the cross-sectional sample as in the original sample. This and the lack of good representation in the cross-sectional sample for most of the range of enlisted paygrades (only paygrades E-3, E-4, and E-5 are well represented) account for more of the difference between the significance of the combination of variables for the two models.

PERFORMANCE EVALUATION MODEL

An objective of the performance evaluation model was to determine which, if any, of the on-the-job evaluation variables were significant when introduced into the model. How do the evaluation variables modify the model? Another objective of this model, of course, was to evaluate the relative impact of the current performance evaluations on minorities' and nonminorities' promotion rates. Furthermore, comparisons were to be made between the advancement function derived from the pre-

---

[20] The cross-sectional model R^2 is .3850. R^2 for the original model is .8059. The cross-sectional model statistics are found in Appendix E (Table E-1).

service plus in-service sample and the advancement function derived from the individuals of this performance evaluation sample.

Only the performance evaluations for E-5 and above are filed on tape. Therefore, the individuals selected for the sample were in paygrades E-5 and above. The variables that were considered for the preservice plus in-service model remained as candidates in this model. In addition, current performance, appearance, cooperativeness, reliability, conduct, resourcefulness, leadership, equal opportunity, and overall evaluations were entered into the model.[21]

Only paygrades E-5 and E-6 are well represented in the performance evaluation model, whereas paygrades E-1 through E-6 are well represented in the original preservice plus in-service model. Furthermore, the performance evaluation model describes the enlisted advancement function during a time frame different from those of either the original model or the cross-sectional model. The original and cross-sectional models have a median value for time in the Navy of twenty-one months and four years, respectively. The performance evaluation model range for time in service is twenty-four months to fifteen years. In this model, the mean and median values for time in the Navy are both seventy-one months. Thus, we are looking not only at a more specific paygrade group than in the other models but also at a different time frame.

Performance Model Interpretation

Of the in-service variables, time in service is the most significant in the model. It does not, however, display as strong a relationship with the dependent variable in this model as it does in the original preservice plus in-service model. On the other hand, the occupational classification variables maintain their relative strength of relationship with the dependent variable in the performance evaluation model.

In the original model, the academic variables were found to contribute, in large part, to the explanation of the dependent variable's variation. ARI and GCT scores were found to be the most significant variables and are very important in determining an individual's opportunity to advance in rank. Years of education and the AFQT were found to be less important,

---

[21] The performance evaluation model R^2 is .4804. The performance model statistics are found in Appendix F (Table F-1).

Statistical Analysis

and CLER and SHOP were found to be far less important, yet statistically significant.

In the performance evaluation model, the academic variables display less joint significance than that displayed in the original model. ARI shows a strong relationship with the dependent variable. SHOP and GCT are the only other academic variables that are statistically significant. These variables do not have nearly as much impact on advancement opportunity as does ARI.

The average BTB and AFQT scores are higher, and the number of years of education completed are greater for personnel in the E-5 and above sample than for members of the original sample. These academic credentials differences are reminiscent of these same variable differences found in the last section between the cross-sectional sample and the original sample. Because of the disproportionate attrition over time of individuals with lower academic credentials, the mean values of these variables increase for those remaining in the sample.

Approximately 4.2 percent of the sample of personnel in paygrades E-5 and above are black, and 1.6 percent are other minorities. In the original model, 10.6 percent and 7.8 percent of the sample are black and other minorities, respectively. The attrition rate for minorities becomes disproportionately high as the median time in the Navy for the sample increases. It is concluded that most of the minorities entering the Navy do not remain in the service for a longer period (relative to the time frame of personnel in paygrades E-5 or greater) or advance to E-5 or greater. (The problem of disproportionate minority personnel attrition is discussed in chapter VII.) Those minority members who do remain, however, enjoy as fast a promotion rate as that of majority members with the same significant variable characteristics. Therefore, neither the black personnel nor the other minority personnel variables are significant in the performance evaluation model.

The joint influence of the current performance evaluations is less than anticipated. Only leadership and appearance show a statistically significant relationship with the dependent variable. Recall, however, that this model describes the advancement function for paygrades E-5 and above only. It is quite possible that these results do not well represent the state of affairs for paygrades E-4 and below.

The model results show that minorities do receive slightly lower scores on the current performance evaluation than do nonminorities. The proportion of both the leadership and appearance variables' variation explained by the race-ethnic variables is, however, less than 1 percent. This indicates that, in the aggregate, there is little difference between race-ethnic groups' scores on leadership and appearance evaluations. This conclusion holds only when referring to group comparisons for paygrades E-5 and above. As stated, no comment can be made here concerning group performance comparisons for personnel in paygrades E-1 through E-4.

SUMMARY

The purpose for conducting this statistical analysis was twofold. First, it was hoped that a good enlisted advancement function could be constructed to explain paygrade level attainment. The second purpose was to evaluate minorities' vs. nonminorities' advancement opportunities with respect to the statistically significant variables derived from the models. On both accounts, the analysis was successful.

Generally, the variables found to be significant are consistent from model to model. Exceptions to this were noted. Academic credentials are important and influence an individual's advancement potential throughout his career. These variables' frequency distributions are strikingly different for majority, black, and other minority personnel. The academic variables account for much of the race-ethnic promotion rate differences. The following in-service variables are statistically related to paygrade level: time in the service, discipline record, leadership and appearance evaluations, and occupational classification. These factors are extremely important to an individual's promotion success, and minorities are adversely affected by the influence of these variables on the advancement function.

In the next chapters, the services' formal advancement and professional development systems are investigated. Attention is turned to the areas of occupational classification and assignment, advancement, and retention to evaluate whether or not the services' institutional policies and affirmative action objectives are compatible. For these discussions, the conclusions derived from the statistical analysis will be helpful in two ways. First, the analysis can answer, in part, the question of the compatibility of affirmative action and institutional policy. Second, the statistical results provide insight into what the effects will be of modifying the policy-regulated variables to accommodate affirmative action objectives.

CHAPTER V

Occupational Classification and Assignment

The military is currently thought by many to do a good job of upgrading the skills and educational levels of those disadvantaged members of our society who join. Adam Yarmolinsky stated, in his study of the military establishment, that

> military service assists certain groups not only through specialized training but even more importantly through the influence of the military as a social organization. . . . The more depressed the socioeconomic background of the recruit, the more he seems likely to benefit from experience in the armed services. . . . Precisely because it is a user of unskilled manpower, the military serves as an agency of basic education. . . . A task is found for each person rather than a person for each task. In this sense, it is a "total" institution, because once admitted, all men have a place. . . . For persons from the lowest social strata, there is a built-in process of upgrading.[1]

Yarmolinsky's picture of the military as a social organization is not quite accurate. It is true that, for a disadvantaged person who enters the military and is retained, the pay, benefits, and security are probably as good or better than what that same unskilled, poorly educated individual could get in the private sector. The military, however, has not lived up to its full potential as a social organization. Because of the aptitude-testing system used to match individuals with jobs, disadvantaged personnel entering the service have been relegated to soft-skill jobs, for which there is little or no formal training. Personnel entering the military with the highest academic credentials are assigned to formal school training and afterward to a technical occupation and are upgraded to a greater extent than the educationally disadvantaged who do not qualify for formal schooling. For individuals assigned to soft-skill jobs, there is limited educational benefit to be gained from the military; furthermore, their

---

[1] For full quotation and additional information, see Adam Yarmolinsky, *The Military Establishment—Its Impacts on American Society* (New York: Harper & Row, 1971), pp. 324, 325.

postservice employment prospects are more limited if they do not learn a technical skill while in the services.

As the largest vocational training institution in the United States, the military has the potential to contribute significantly toward the attainment of national manpower goals. A major goal of national manpower policy is to improve the distribution of employment opportunities and career prospects for those individuals facing labor market barriers [2] caused by those individuals' lack of job skills, deficiencies in basic education, etc.

ENLISTED MINORITY PERSONNEL OCCUPATIONAL DISTRIBUTION

The services have moved quickly over the past seven years in an attempt to ensure that the percentage of minorities within their ranks equals the percentage of minorities in the United States' population. In 1970, 5.4 percent of the enlisted Navy was black. As of June 1977, 8.5 percent of the Navy was black. During the same period, the Marine Corps increased its black participation rate from 11.2 percent to 17.3 percent.[3] There is no question that both the Navy and the Marine Corps have made significant strides in recruiting proportionate numbers of minority personnel.

Another vital prerequisite, however, is the dispersion of that participation. The distribution, or frequency, of enlisted minority personnel throughout the hierarchy of occupational skill levels directly reflects the degree to which minorities participate in the more highly skilled and technical job categories. The frequency distribution of Navy and Marine Corps enlisted men by race and Department of Defense Occupational Group [4] is provided in Table V-1.

In the Navy, blacks are overrepresented, relative to the percentage of total personnel assigned to a particular occupational

---

[2] National Commission for Manpower Policy, "The Quest for a National Manpower Policy Framework," Special Report No. 8 (Washington, D.C.: U.S. Government Printing Office, 1976), pp. 1-20.

[3] Defense Manpower Data Center, Military Master Files as of July 30, 1978.

[4] The Department of Defense has developed ten enlisted occupational groups into which it has classified all military enlisted specialties. This serves to somewhat standardize occupations between services for reference. See Office of Assistant Secretary of Defense, Manpower, Reserve Affairs and Logistics, *Occupational Conversion Manual* (Washington, D.C.: Department of Defense, 1977).

Occupational Classification and Assignment 97

TABLE V-I
Percentage Distribution of Enlisted Men by Race and
Occupational Group
in the Navy and Marine Corps
December 31, 1977

Department of Defense Occupational Group	Navy Total	Navy Black	Marine Corps Total	Marine Corps Black
Infantry, Gun Crews and Shipboard Specialties	9.7	10.4	30.3	38.9
Electronics Equipment Specialists	13.1	7.1	8.0	2.1
Communications and Intelligence Specialists (includes Radar and Air Traffic Control)	9.5	11.3	7.9	6.2
Medical and Dental Specialists	5.1	5.4	a	a
Other Technical and Allied Specialties (includes Photography, Drafting, Surveying, Mapping, etc.)	1.8	1.1	1.6	1.3
Administrative Specialists and Clerks	14.2	19.9	16.8	16.7
Electrical/Mechanical Equipment Repairmen	31.3	26.6	15.3	9.4
Craftsmen	6.0	3.7	3.2	3.0
Service and Supply Handlers	5.8	8.3	14.8	19.7
Miscellaneous Others	3.5	6.2	2.1	2.7
TOTAL	100.0	100.0	100.0	100.0

Sources: Bureau of Naval Personnel, "Enlisted Personnel by Grade, Years Active Service, Mental Groups and Racial Groups," MAPMIS 5314-4108 (Washington, D.C.: Department of the Navy, February 23, 1978); U.S. Marine Corps, Headquarters, Manpower Planning, Programming and Budgeting Branch, August 1978.
[a] Serviced by the Navy in this occupational group.

category, in the infantry, gun crews and shipboard specialties; communications and intelligence occupations; administrative specialists and clerks category; service and supply handling category; and miscellaneous functions. They are underrepresented as electronics equipment specialists, electrical and mechanical equipment repair personnel, other technical and allied personnel, and craftsmen. Roughly proportionate representation exists in the medical and dental specialties.

The principal differences between the total personnel and black personnel percentage distributions in the Navy are found in the following occupations: miscellaneous others, electrical/mechanical equipment repair, and electronics equipment. In the least skilled areas categorized as miscellaneous, 6.2 percent of all enlisted black, but only 3.5 percent of the total enlisted population, are members. The miscellaneous others category consists largely of enlisted personnel in paygrades E-3 and below who have not been sent to formal technical training school and, therefore, do not have an occupational classification. On the other hand, in the highly technical fields such as electrical/mechanical equipment repair and the electronics equipment specialties, a much smaller percentage of blacks are working in these areas than the percentage of the total enlisted force assigned to these fields. For example, 7.1 percent of blacks are working as electronics equipment specialists. This is compared with 13.1 percent of all enlistees found in the field.

Generally, black marines are more evenly distributed across job categories than are blacks in the Navy. In the Marine Corps, blacks are proportionately represented as communications and intelligence specialists, other technical and allied specialists, administrative specialists and clerks, craftsmen, and miscellaneous others. In the least skilled categories—which include infantry, gun crews and shipboard specialties; service and supply handling; and miscellaneous others—the total manpower allocation is approximately 47 percent. Yet, 61 percent of blacks find themselves in these categories. In the most technical of job categories (electronics equipment specialists), only 2.1 percent of blacks are participating, whereas the participation rate for all enlisted men is 8.0 percent. Blacks are also underrepresented as electrical/mechanical equipment repairmen in the Marine Corps. Both are categories in which minorities are underrepresented in industry.

Progress has been made in recent years to distribute minorities proportionately across occupations. Yet much of the work needed to solve the tougher problem of proportionate occupational distribution remains. Historically, minorities have been relegated to the so-called soft-skill occupations. Without affirmative action measures, the services cannot hope to achieve a more balanced composition within the foreseeable future.

THE CLASSIFICATION AND ASSIGNMENT PROCESS

In the prevolunteer military, the recruit assignment authority was almost entirely held by the classification branch of the service.[5] Operationally, few assignment prerogatives were granted to prospective recruits. Basically, formal training school assignment and subsequent occupational classification consisted of screening an individual with paper-and-pencil mental aptitude exams to determine which formal schools and, therefore, which jobs the individual was eligible for.[6] Then, the service selected an occupation for the recruit based on service-established priorities and the individual's aptitudes.

The draft guaranteed a continual supply of "draft-induced" volunteers. The draft-guaranteed supply ensured a favorable personnel quality balance for the military. The quality of recruits was consistently high. There was no need to offer inducements such as guaranteed formal assignments to attract better qualified individuals.

Once an individual was recruited into the service, he was sent to recruit training, or boot camp. Early in the initial training phase, the recruit was administered a battery of standardized tests. For the majority of recruits who were not under a guaranteed assignment contract, the examination battery did, to a large extent, determine the set of occupational specialties for which the recruit was eligible.

The classification process also included the use of personal interviews with recruiters and/or occupational counselors, intensive background checks for some occupations, and physical examinations. Importantly, for most recruits, the classification process, including mental aptitude testing, took place at boot camp after the individual had joined the service. Those recruits who were guaranteed a formal school seat at enlistment were required to demonstrate their mental aptitude qualifications

---

[5] Much of what is discussed in this section is taken from a report prepared by the Defense Research Projects Agency. Fred Morgan and Darien Roseen, *Recruiting, Classification and Assignment in the All-Volunteer Force: Underlying Influences and Emerging Issues*, R-1357-ARPA (Santa Monica, Calif.: The Rand Corporation, 1974).

[6] Formal school training is generally required for entry into technically oriented or skilled occupational specialties. Those who are not qualified for formal school training go directly to their operational units as undesignated strikers.

through the Armed Forces Qualification Test (AFQT). The AFQT was administered at the recruiting station.

Assignment decisions were based on many factors. Available in-fleet positions, training program openings, previous commitments made by recruiters,[7] transportation costs, aptitude exams, and staffing priorities are factors that were, and still are, considered when assignment decisions were made.[8] The hierarchy of prevolunteer force assignment objectives is summarized in Figure V-1.

Officially, the honoring of recruiting commitments was given highest priority in the assignment of trainees. What is not often remembered, however, is that, relative to the all-volunteer force, few recruiting commitments were given. Furthermore, when given, the commitments were generally so broad that they did not interfere with other objectives. Operationally, then, the most influential assignment objective was to fill the available formal service school openings.[9] The prevolunteer assignment process, therefore, was "a sequence of screening operations which identified those jobs which a recruit was eligible to hold and then selected one job from among the eligibles primarily on the basis of service established priorities and the recruit's aptitudes." [10]

Volunteer Assignment

Modifications to the classification and assignment process have been made in an effort to increase the attractiveness of service in the all-volunteer Navy and Marine Corps. One such modification calls for a larger percentage of recruits to enter the service with some form of commitment which must be honored.[11] Also, mental aptitude exams are now administered at the re-

---

[7] Guaranteed formal school training is one of several possible types of commitments that can be made at recruitment. Recruits who are given a formal school commitment at enlistment have already passed the minimum qualifications for that occupational area or specialty. Other commitments include at-sea options, preferred geographic location, etc.

[8] Morgan and Roseen, *Recruiting, Classification and Assignment*, p. 18.

[9] Assignment to a formal school implies subsequent assignment to a related job.

[10] Morgan and Roseen, *Recruiting, Classification and Assignment*, p. 20.

[11] Defense Manpower Commission, *Defense Manpower: The Keystone of National Security* (Washington, D.C.: U.S. Government Printing Office, 1976), p. 191.

FIGURE V-1
Prevolunteer Service Hierarchy of Assignment Objectives

---

1. Honor recruiting commitments,
    Then
2. Fill schools in order of priority assigned to each by service,
    Then
3. Whenever possible, fill each school without violating minimum qualification standards (physical, mental, etc.),
    Then
4. Minimize transportation costs of school assignments (Army and Navy only),
    Then
5. Maximize preferences and/or counselor recommendations,
    Then
6. Maximize aptitudes of those assigned to classes,
    Then
7. Assign remaining trainees to directed duty assignments in accordance with service priorities, transportation costs, preferences, recommendations, in that order.

Source: Fred Morgan and Darien Roseen, *Recruiting, Classification and Assignment in the All-Volunteer Force: Underlying Influences and Emerging Issues*, R-1357-ARPA (Santa Monica, Calif.: The Rand Corporation, 1974), p. 19.

cruiting stage rather than at boot camp. This provides each recruit with the opportunity to know before enlisting which programs he is eligible to participate in. In order to qualify for a particular occupation requiring formal schooling, the recruit must still meet the eligibility criteria.

The all-volunteer Navy and Marine Corps have maintained the assignment priorities outlined in Figure V-1. As a result of the recruiting realities of the all-volunteer force, eligible recruits now take more of a part in the assignment decision as a condition of enlistment. If a recruit is qualified, the assignment decision often consists of that recruit's selecting an occupation from among the list for which he qualifies.

Thus, classification has shown some movement away from its traditional focus on aptitudes. This shift, however, only affects man/job matches when the individual can satisfy the criteria for a particular formal school and related job. Yet, in classification decisions, a greater emphasis is beginning to be placed on the recruit's occupational preferences. This is a major deviation from the stricter aptitude-matching philosophy of earlier years.

Minorities, too, are now given more of an opportunity to determine which occupational fields they will enter. If an individual is not satisfied with the arrangements, he can decide not to enlist. Mental aptitude requirements for formal school training remain. These requirements are the greatest deterrent to full participation by minority personnel in the technical occupational specialties. Today, however, individuals who do not qualify for formal training are made aware of this before enlisting. Therefore, it is less likely that an individual's expectations will be suddenly diminished once he enlists.

Enlistment Options vs. Open Contract Assignment

The Navy and Marine Corps are increasingly using enlistment options and assignment guarantees as inducements to join. This has been motivated by the all-volunteer force and the consequent strong competition for highly qualified individuals between the civilian and military institutions (and between the services) in the manpower marketplace.

In the Navy, for fiscal 1977,[12] between 55 and 60 percent of total recruit accessions were scheduled to be assigned to formal school (Class A school) training. This projected rate was consistent with the fiscal 1976 rate of recruits' formal school participation.[13] Generally, formal school training is a necessity for advancement in the technical or skilled occupational specialties, but approximately 40 to 45 percent of the enlisted recruit population do not attend formal schools. These individuals, so-called undesignated strikers, report directly to their operational commands [14] after boot camp. For the most part, undesignated strikers are relegated to the unskilled occupations where promotion rates are slower.

---

[12] October 10, 1976, to September 30, 1977.

[13] Chief of Naval Operations, "Approved Class 'A' School Training Impact Plan for FY77," Serial 992f5/72724, to Bureau of Naval Personnel, Department of the Navy, Washington, D.C., October 9, 1975.

[14] An operational command is the working environment within which an individual is assigned a job. It is an organization with a mission. Individuals within the command work to accomplish that mission.

Occupational Classification and Assignment

The Recruiting Command is authorized to guarantee up to 90 percent of the available formal school seats at recruitment.[15] Therefore, the remaining school seats, at least 10 percent of the total available, are used for in-fleet distribution. These formal training school assignments are called in-fleet assignments.

In the Marine Corps, for fiscal 1976, approximately 60 percent of the total recruit accessions were enlisted under the provisions of an enlistment guaranteed program.[16] Therefore, approximately 40 percent of the recruits joined the Marine Corps under an open contract provision. Individuals who enlist under an open contract are assigned to an occupational specialty at boot camp. There, the decision is based upon the needs of the service and the individual's aptitude and other qualifications. The recruit's occupational preferences are also considered, but to a lesser degree.

Selection for formal school training is contingent upon an individual's scoring sufficiently well on the aptitude tests. It is not surprising, then, that a disproportionate number of minorities are recruited under an open contract and are assigned to an operational command as undesignated strikers.

The Marine Corps makes use of an Enlistment Bonus Program which is used to assist in attaining adequate numbers of enlistments in designated occupational specialties. The Enlistment Bonus comprises the Combat Arms Enlistment Bonus and the Technical Skills Enlistment Bonus programs. These programs provide a pay incentive for highly qualified individuals, as measured by the mental aptitude exams, to enlist in a particular occupation for at least four years. The Combat Arms Bonus is used to induce highly qualified individuals into the combat arms ratings. These ratings constitute the soft-skill occupations and are normally occupied by the least qualified recruits. The Technical Skills Enlistment Bonus is used to induce highly qualified individuals into the technical ratings in which critical manpower shortages exist.[17]

---

[15] This figure includes those personnel selected for A school after arriving at basic training.

[16] U.S. Marine Corps, Headquarters, "FY1976 Marine Corps Recruit Occupational Classification (MCROC) Statistical Summary" (Washington, D.C., 1976).

[17] U.S. Marine Corps, Headquarters, "Enlistment Bonus Program," Marine Corps Order 1130.57B (Washington, D.C., 1976).

The mental aptitude requirements for the combat arms ratings are relatively low, and minorities are disproportionately represented in these ratings. The Combat Arms Bonus is designed to increase the number of highly qualified individuals in the combat arms ratings by offering cash incentives and to help provide for increased race-ethnic parity in these occupations. It is possible, however, that highly qualified minorities will opt for the Combat Arms Bonus rather than for the Technical Skills Enlistment Bonus because of the cash incentive differences.[18] This could work counter to the Marine Corps' equal opportunity objective of increasing minority personnel participation rates across the more technical occupational specialties. Research is needed to determine what influence the Enlistment Bonus Program is having on minority occupational distribution patterns.

The normal armed services enlistment contract extends for four years. The Navy has a five-year enlistment contract. Both the Navy and the Marine Corps also have a six-year enlistment contract. The five- and six-year enlistment programs are small relative to the four-year program. Individuals who enlist under the six-year contract are assigned to advanced technical training in occupations such as electronics or nuclear power.[19] Eligibility requirements for the six-year enlistment program are particularly stringent. At the completion of formal school training, individuals who enlist under the six-year program are automatically advanced to paygrade E-4.[20] Historically, few minorities have been selected for the six-year program.

In-Fleet Formal Training Assignment

To be eligible for assignment to a formal school directly from boot camp, a recruit must meet the stringent mental aptitude eligibility requirements. It is not uncommon for the mental aptitude requirements to disqualify a majority of recruits who are interested in a particular occupational specialty. Waivers

---

[18] The Combat Arms Enlistment Bonus is a $2,500 cash payment. The Technical Skills Enlistment Bonus is a $1,500 cash payment. U.S. Marine Corps, Headquarters, Marine Corps Order 1130.57B.

[19] Navy Recruiting Command, *Navy Recruiting Manual—Enlisted*, COMNAVCRUITCOM Instruction 1130.8A (Washington, D.C.: Department of the Navy, 1974), p. 26-1.

[20] The enlisted paygrade (rank) range is E-1 to E-9. E-1 is the lowest enlisted paygrade; E-9 is the highest. E-4 is the most junior of the enlisted petty officer rates.

are seldom approved for those recruits aspiring to formal school training directly from boot camp and not meeting the eligibility requirements.

At least 10 percent of the Navy formal training school seats are filled by in-fleet assignments. The in-fleet assignment procedure provides a means by which individuals who have been sent directly to jobs without formal school training may be recycled to formal school. A positive endorsement is required from the commanding officer of the unit to which the individual is attached. When an individual is recommended for formal school training after having worked in an operational command for a period of time, aptitude waivers are generally granted.

Minorities are disproportionately represented in the lower aptitude categories and, therefore, are disproportionately represented in the soft-skill occupations. The in-fleet assignment process provides a mechanism through which those individuals who did not qualify for formal school training from boot camp can be given another opportunity, based upon in-service performance, to receive formal training. A large percentage of the in-fleet selectees are minorities.[21]

The in-fleet formal school assignment process is a tool which should be fully utilized, and possibly expanded, to upgrade individuals who show promise, but lack the education to have qualified for formal school in the aptitude examinations. It is clear that minorities are fully represented in this category. It is one way in which individual commands can directly participate in upgrading and can have a very positive impact on minority utilization.[22]

UNDESIGNATED STRIKERS AND ON-THE-JOB TRAINING

Despite the substantial increase in guaranteed enlistment contracts being given by recruiters and the detailing of recruits at boot camp to fill the balance of openings in formal schools, many recruits are sent directly from boot camp to operational commands. Generally, each operational command is free to assign this personnel to jobs in accordance with the needs of

---

[21] Navy Recruiting Command Program Analysis 1100-3 Series, supplemental sheets, December 31, 1975, June 1976.

[22] Aptitude exams are partly a measure of scholastic achievement. See Arthur I. Siegel, Brian A. Bergman, and Joseph Lambert, *Nonverbal and Culture Fair Performance Prediction Procedures*, Vol. II, *Initial Validation* (Wayne, Pa.: Applied Psychological Services, Inc., 1973), pp. 2-3.

the command. Recruits who are assigned to a command directly from boot camp are categorized as undesignated strikers.[23] They have no formal school occupational training and are normally placed by the operational command into nonskilled occupations in which personnel shortages exist. In the Navy, between 40 and 45 percent of the recruit accessions are assigned to commands as undesignated strikers. In the Marine Corps, the percentage is approximately 40 percent. Minorities are disproportionately represented in the occupations which require no formal school training and in which, therefore, personnel reports directly to operational commands as undesignated strikers.

Designated strikers, on the other hand, are those individuals who, because of their formal school training or special skills and aptitudes, are sent to their first operational command as apprentices in the rating that corresponds to their training or skill. These occupations generally have faster promotion rates.

Undesignated Striker Job Selection Process

The Navy and Marine Corps have similar policies on the process of job selection for the undesignated striker. Because both services have similar policies, only the Navy procedure for assigning an undesignated striker to an occupation is discussed here. A general set of guidelines exists for the Navy striker procedure,[24] but its application varies greatly from one command to the next.

When a sailor just out of boot camp checks on board a ship for duty, his first few days (normally between one and five workdays) are spent on his becoming oriented to the ship: its operations, organization, physical structure, as well as the services which are available (medical, dental, educational, legal, etc.). After this brief introductory period, he is usually assigned to work in the galley or deck force for a period of up to ninety days' ship's service work. It is generally a period of adjustment for the sailor, both to real Navy life and to his new environment. During this ninety-day period, the sailor is expected, and usually has the opportunity, to visit the various functional areas of the ship, to learn about the jobs which are per-

---

[23] Strikers, either designated or undesignated, are in apprenticeship at paygrades E-1, E-2, or E-3.

[24] Bureau of Naval Personnel, *Bureau of Naval Personnel Manual*, NAVPERS 15791B (Washington, D.C.: Department of the Navy, 1969), Article 2230220.

formed there, and to develop an interest in one or more of these occupational areas.

Near the end of his ship's service work assignment, the recruit submits a list of occupational preferences to a "striker board," which consists of senior enlisted men and officers from the command of different ranks and different occupational fields. After considering the individual's request for permanent assignment and the manpower needs of the ship, the striker board assigns him a particular occupation in which he begins on-the-job training.

Problems with Job Selection and Training

The preceding is a general and ideal description of what the striker procedure is like in the Navy fleet. Similar procedures exist for other operational units of the Navy and for the Marine Corps. Individual commands are tasked with working out the *specific* procedures to be followed in assigning an undesignated striker to a particular occupation. Because of this, the "ideal" is often not applied. Each command has considerable latitude in this area. Some commands do not have a well-delineated assignment procedure, and even within a single command, a variety of procedures may exist.[25]

A common subject of complaint among individuals who are sent to commands as undesignated strikers is the ship's service-work phase of their enlistment. The period of ninety days' ship's service work is often looked upon as a degrading and demoralizing experience. Most recruits were not informed at recruitment of this portion of Navy life, and this fact merely reinforces the frustration involved. A frequent response from the undesignated striker is "if I had wanted to be a janitor, I could have worked in my hometown."

Many senior officers and enlisted personnel see the young undesignated recruit as lacking motivation to perform well even though his duties are menial. If a lack of motivation does exist, however, it may not be inherent in the individual, but may have been caused by situations in which the individual's motivation to achieve has been frustrated by the organization.[26]

---

[25] On-board ship interviews were conducted at the naval port facilities of Philadelphia, Norfolk, and San Diego, 1973-1976.

[26] D. E. Super, *The Psychology of Careers* (New York: Harper & Row, 1957); and Russell Doré and Merle Meacham, "Self Concept and Interests

Many of the individuals sent to an operational command without formal schooling unwillingly remain in such unskilled jobs as cook or deck force personnel. They are utilized in accordance with the needs of the command. Moreover, in many cases, information concerning occupational openings is not well publicized. This is one area in which considerable improvement can be made in providing minorities with the opportunity to upgrade themselves.

Recommendations for Upgrading Striker Placement Opportunities

The Navy and Marine Corps should establish specific guidelines for a dynamic system of personnel assignment to on-the-job training. Such a system should include a complete orientation of new personnel to the command, counseling on job opportunities available, and the opportunity actually to work in areas of interest prior to making a choice.

Command occupational shortages and anticipated shortages should be widely publicized. The undesignated striker should be made aware of the situation and allowed to vie for the job through the striker board. If the individual has proved to the command's satisfaction that he is motivated and capable, then he should be provided with an opportunity to enter the desired occupation. A mechanism which provides real occupational alternatives and opportunities to those motivated and capable undesignated strikers is essential if the services are to achieve proportionate occupational distribution of minorities.

An undesignated striker who is given an opportunity to enter an occupational specialty which normally requires formal training must be allowed to attend formal schooling. In the more skilled and technically oriented fields, formal schooling is a must

---

Related to Job Satisfaction of Managers," *Personnel Psychology*, Vol. 26 (1973), pp. 49-59.

Doré and Meacham draw on Super's theory that the career development process implements a self-concept. Super states that a job serves to implement self-concept, and that when the self-concept and the job do not match, dissatisfaction results. From an applied viewpoint, Doré's and Meacham's study provides some valid predictive instruments for use in helping to predict job satisfaction when counseling people who are considering management careers. The military should seriously consider incorporating a methodology which measures an individual's job needs (vocational interest, etc.) into their enlisted personnel classification composite. The above researchers have found that use of such a device would increase the probability of a suitable occupation/personnel match and of an individual's job satisfaction and, possibly, productivity.

for successful advancement in rate. Passing the advancement exams for the next higher paygrade is a prerequisite for promotion in all occupations. In the more technically oriented specialties, knowledge received from on-the-job training is not enough to pass the examination. An understanding of theory is required, which is best acquired at formal schools.

Individuals within an occupational specialty who have formal school training are clearly at a competitive advantage in terms of promotion rate. Therefore, processes for recycling for formal training those recruits who reported to their commands without formal training and who show promise must be fully utilized.

Industry Practices

In attempting to compare and contrast the manpower practices of the services with those of private industry, there are immediate constraints which render the two institutions incomparable. A common practice in private industry is to give the prospective employee a tour of the plant and to familiarize him with the job he will be doing and the people with whom he will be working. The individual's decision to accept a job will be based on his evaluation of the job and the working environment. Naturally, such procedures could not be applied to the services for numerous reasons, principal among which is the broad geographical dispersion of their operations.

There are at least two practices which private industry employs, however, that the military services might incorporate into their manpower policies. The first would approach the problem of recruit motivation. In making an employment offer, private industry advises the individual of all conditions of his employment—the job, the people, the organization and its functions, the benefits, and more. Some of these conditions are categorically impossible for the services to present, but the underlying principle of thorough information on the job and the working environment is necessary for recruits to maintain their motivation and to work within the military organization toward the realization of their goals.

A second practice involves the flow of communications concerning job openings. Many companies post job openings on special bulletin boards throughout the firm. All employees are then given the chance to see what is available within the organization for their possible upgrading and/or lateral transfer. The services could adopt such procedures at the command level for their on-the-job training openings. It would make the new sailor or marine immediately aware of command needs and of where his interests might be most profitably applied.

FORMAL TRAINING SELECTION AND ASSIGNMENT

In order to qualify for particular occupational specialties and schools, recruits must meet medical and security requirements, as well as mental aptitude qualifications. The mental standard requirement is, operationally, the most stringent and pervasive formal school [27] selection criterion.

Selection for Formal School Training

Each of the armed services uses a series of mental aptitude test batteries designed to estimate the potential success of individuals in training for various specialties. Prior to January 1976, recruits who did not enlist under a guaranteed formal school option took the Basic Test Battery (BTB). The BTB was administered during the initial phase of boot camp. The BTB is an examination battery consisting of six sections, which are described in Figure V-2. The examination battery was combined according to specific formulas to yield selector scores for occupational fields. For those recruits who were not under a guaranteed assignment contract, the selector scores did, to a large extent, determine the set of occupational groupings for which they were eligible. Those who enlisted under a guaranteed assignment contract fulfilled the formal school eligibility requirement during the recruitment stage by doing sufficiently well on the AFQT.

In January 1976, the Armed Services Vocational Aptitude Battery (ASVAB) replaced the BTB as the primary test used for enlistment screening and classification.[28] Both the ASVAB and the BTB ostensibly provide accurate information on an individual's mental aptitude and vocational aptitude. In most instances, the ASVAB qualifications for a particular school or program can be readily ascertained by matching the BTB sub-

---

[27] Formal school training is designed to provide the recruit with a basic theoretical and practical knowledge of a particular occupational specialty. Upon graduation from formal school, the recruit is ordinarily assigned to a job in that occupational specialty.

[28] Bureau of Naval Personnel, NAVPERS 15791B, Article 1440220.

FIGURE V-2
The Basic Test Battery (BTB)

The *General Classification Test (GCT)* is a measure of ability to comprehend and define words and to reason verbally. It is represented most heavily in what is often termed as reading skill. Vocabulary is only a factor which characterizes reading skill; but it provides a measure of verbal comprehension.

The *Arithmetic Reasoning Test (ARI)* is a test of quantitative aptitude involving mathematical reasoning and problem solving.

The *Mechanical Comprehension Test (MECH)* measures aptitude for mechanical work, mechanical and electrical knowledge, and the ability to understand mechanical principles.

The *Clerical Test (CLER)* measures the ability to observe details rapidly and measures the speed of responses to observations.

The *Shop Practices Test (SHOP)* measures the functional ability of an individual who has had experience with, and is knowledgeable about, the use of a variety of tools found in a shop.

The *Electronics Technician Selection Test (ETST)* measures knowledge of mathematics, science, electricity, and electronics.

Sources: Bureau of Naval Personnel, *Bureau of Naval Personnel Manual*, NAVPERS 15791B (Washington, D.C.: Department of the Navy, 1969), Article 1440220; and idem, *Manual of Enlisted Classification Procedures*, NAVPERS 15812B (Washington, D.C.: Department of the Navy, 1970).

tests with the ASVAB equivalents. The ASVAB series compares with all subtests on the BTB as follows:[29]

BTB	ASVAB
General Classification Test (GCT)	Work Knowledge (WK)
Arithmetic Reasoning Test (ARI)	Arithmetic Reasoning (AR)
Mechanical Comprehension Test (MECH)	Mechanical Comprehension (MC)
Clerical Test (CLER) or Coding Speed Test (CST)	Numerical Operations (NO) and Attention to Detail (AD)
Shop Practices Test (SHOP)	Shop Information (SI)
Electronics Technician Selection Test (ETST) or Electronics Selection Test (EST)	Electronics Information (EI), Mathematics Knowledge (MK), and General Science (GS)

[29] *Ibid.*

The examination subtests are combined by formula to provide selector scores used for formal school placement. The subtest combination deemed most appropriate for the reliable prediction of successful completion of a particular formal training is used. For a Navy enlistee to be considered for formal school training, he must have fulfilled one of the following requirements:

1. an AFQT score of 49 or greater, or

2. an ASVAB composite score of WK + AR = 100 or greater, or

3. a combined GCT + ARI score of 100 or greater on the BTB, or

4. a combined GCT + ARI + MECH of 148 or greater on the BTB, or

5. qualification for, and enlistment in, a program that guarantees formal school training by meeting the BTB (if tested prior to January 1, 1976) or ASVAB score requirements.[30]

The more technical schools and occupations require higher selector scores.

Aside from the series of mental aptitude exams, the classification process for the recruit includes personal interviews with recruiters and/or occupational counselors and a medical examination. The eligibility requirements for formal school training in an occupational specialty include more than just satisfying mental aptitude requirements. Specific medical and moral standards are established and enforced as well, which serve as potential disqualifiers. An individual's emotional stability, for example, is taken into consideration during the classification process. A recruit can also be disqualified from further consideration for reasons such as a criminal record or drug abuse. For some occupations, intensive personnel background checks are required before an individual is selected.

Aptitude Testing: From AFQT and BTB to ASVAB

The ASVAB is a joint armed services test administered at the Armed Forces Examining and Entrance Station. It is not

---

[30] Assistant Chief of Personnel Planning and Programming (Pers-2123/cm), "Recruiting Goals and Policies," Serial 797/76, Memorandum for Navy Recruiting Command, Department of the Navy, Washington, D.C., 1976.

Occupational Classification and Assignment

administered by recruiters but rather by a joint services committee. Development began in 1974 to provide each service with aptitude measurement areas comparable to their current test batteries and, so, to achieve standardization of mental testing at the enlistment point.[31] As already stated, ASVAB became operational in January 1976.

Prior to ASVAB, the Navy and Marine Corps used either the AFQT or the short version of the Basic Test Battery to determine enlistment eligibility and, increasingly, occupational classification. The exams were administered by the recruiter. The recruiter allocated his guaranteed formal school seats to individuals meeting the mental category eligibility requirements at recruitment. The BTB was then administered at boot camp to determine job classification for those not guaranteed formal school training. All recruits were required to take this additional battery of tests regardless of whether or not they had already been guaranteed specialized job-skill training after boot camp as a part of their enlistment contract. It was not uncommon for recruits with guaranteed assignments to score significantly lower on the examination battery administered at recruit training than they did in the recruiter's office.

One hypothesis for the above phenomenon is that some recruiters, because of heavy institutional pressures such as the quota system, "helped" potential enlistees to score well enough for a job skill guarantee.[32] The contractual arrangement of the guaranteed assignment program allowed the recruit to insist on being sent to the school cited in the contract despite the high probability of failure. When this took place, it was not uncommon to find the recruit remaining in the general occupational field, but succeeding only in the lower skill level jobs within the broad occupational group.[33] The alternative was for the recruit to sign a new contract for occupational training at a level for which he was qualified. Minorities have almost certainly felt the implications of the above situation to a greater extent than

---

[31] Bureau of Naval Personnel, NAVPERS 15791B, Article 1440220.

[32] Defense Manpower Commission, *Defense Manpower: The Keystone of National Security*, p. 195.

[33] If a recruit fails to meet the aptitude eligibility requirements for a formal school, that individual will almost certainly have trouble passing the paper-and-pencil advancement exams required for promotion in that occupational specialty.

nonminorities, for the former have traditionally experienced the severest testing problems.

The new ASVAB system precludes the recruit from being "helped" by recruiters on the aptitude examinations. All potential recruits now take the exam under the same test conditions. Furthermore, with the services' increasing their percentage of recruits who are guaranteed formal school training, the ASVAB system negates the necessity of duplicate testing at the recruitment stage and at boot camp. The ASVAB is used to determine both enlistment eligibility and occupational classification.

Minority Personnel Qualifications: The Problem of Academic Credentials

The services' academic requirements have historically been a function of the needs of the service and labor market conditions. The Navy and Marine Corps have set their goals higher for recruitment eligibility for fiscal years 1977 and 1978. The Navy's goal for both years is to recruit 82 percent high school graduates, with 76 percent eligible to attend formal school.[34] The Marine Corps recruited 67 percent high school graduates during fiscal 1976 and planned to increase this figure to 75 percent by the end of fiscal 1977.[35] These higher academic credentials goals, which the Navy and Marine Corps are instituting, will exclude a disproportionate number of minorities from recruitment consideration.

The Navy has categorized job occupations into three groups based upon promotion opportunities within each rating. The three groups are open, neutral, and closed ratings.[36] Minorities are severely underrepresented in the occupations with the greatest advancement potential.[37] All of the occupations in the open

[34] Assistant Chief of Personnel Planning and Programming (Pers-2123), "Recruiting Goals and Policies," Serial 1241/77, Memorandum for Commander, Navy Recruiting Command, Department of the Navy, Washington, D.C., 1977.

[35] U.S. Marine Corps, Headquarters, "Report on Marine Corps Manpower Quality and Force Structure" (Washington, D.C., 1975), p. 17.

[36] Bureau of Naval Personnel, "Career Reenlistment Objectives (CREO)," BUPERS Instruction 1133.25C (Washington, D.C.: Department of the Navy, 1975).

[37] With respect to the sample used to generate the advancement model, 42 percent of total personnel were assigned to an open rating; however, only 27 percent of the blacks in the sample were assigned to an open occupational specialty.

group require formal school training, and the majority are technical and require GCT + ARI equivalents well in excess of 100.[38]

Entrance into the highly technical specialties, such as advanced electronics, requires an individual to be categorized as mental group I or mental group II.[39] Of 1977 accessions in the Navy, less than 3 percent of mental group I and less than 7 percent of mental group II were minorities. Minority accessions were slightly better represented in these categories in the Marine Corps, with approximately 4 and 10 percent, respectively.[40] As you can see, of 1977 accessions, relatively few minorities were deemed qualified for the highly technical occupations.

The selector scores derived from the paper-and-pencil aptitude examinations are of paramount importance in determining formal school, as well as occupational specialty, eligibility. It is easy to see that the academic requirements for formal school eligibility are the major barrier to proportionate minority representation across occupational specialties.

MINORITY PERSONNEL UPGRADING VS. APTITUDE TESTING

The Navy and Marine Corps classification philosophy, as now practiced, emphasizes mental standards. Both services have found it convenient to stress mental standards for two interrelated reasons. First, occupations within the Navy and the Marine Corps have become increasingly more technical and specialized. Individual occupational specialization has become a necessity. The services try to match individuals with occupational specialties through mental categorization so that an individual's success in attaining the required job skills is ensured within a statistically significant range. Second, both services have historically been subject to high turnover rates. The Navy and Marine Corps, therefore, are quite concerned with providing for an individual's success in military training within a prescribed time period. The rapid training of individuals makes

---

[38] Bureau of Naval Personnel, "Selection of Recruits for A School Training," BUPERS Instruction 1236.4 (Washington, D.C.: Department of the Navy, 1976).

[39] *Ibid.*

[40] Bureau of Naval Personnel and U.S. Marine Corps, Headquarters, figures. See Tables III-2, III-3.

possible longer active field service during an enlistment period. Thus, the time constraint has further involved the use of mental categorization.

Mental Aptitude Inflation

The philosophy of rapid training held by the armed services and the implicit aptitude requirements imposed by such a philosophy have acted to inflate the aptitude criteria for occupational specialty training. The available manpower pool has also acted to keep aptitude standards artificially high. One of the chief influences in establishing aptitude requirements for occupational specialties has been the quality of available manpower. That is, the aptitude criteria are somewhat arbitrary. The quality available has tended to become the quality required.

The requirements for entry into a particular occupational specialty have been influenced, to some extent, by the needs of the training commands for students with reasonable probabilities of passing technical training courses. It is by no means clear that the cut-off score used today on the examination to determine formal training eligibility provides an accurate picture of the level which personnel incapable of satisfactory job performance will score below. A high score on the BTB or ASVAB examinations is not necessarily a good predictor of successful job performance.

Should the Navy and Marine Corps adhere to existing quality standards? By imposing high, possibly inflated aptitude requirements, only the individuals with the highest aptitudes will be selected for formal school training. Research indicates that persons with higher aptitudes are more likely to complete formal training. It is argued, therefore, that the training of higher aptitude personnel is generally less costly than providing for formal training of lower aptitude personnel. This argument, however, ignores important questions such as How much less costly? and What are the relative retention probabilities of higher aptitude and lower aptitude personnel?

Certainly there are many new factors which must be taken into account as a result of the all-volunteer force. For example, what effect does the all-volunteer force have on retention rates? Is the services' selection-for-training philosophy, as it stands today, the most cost-effective? Research must be conducted to determine whether selection of only the highest aptitude individuals is really less costly overall than modification downward of the formal school selection scores to allow lower aptitude individuals to participate. If persons with lower aptitudes generally have a better retention rate than those with higher aptitudes and can successfully complete formal school training, then perhaps overall costs can be lowered by adjusting formal school selection scores downward. This notion is certainly worthy of careful analysis and experimentation.

Aptitude Test Validation

The question of aptitude tests' ability to predict successfully an individual's performance on the job is being closely investigated by both the Equal Employment Opportunity Commission (EEOC) and the Office of Federal Contract Compliance Programs (OFCCP).[41] These agencies are concerned with whether racial discrimination in either job placement or advancement is taking place. Because of EEOC and OFCCP scrutiny, the use of testing is declining in American business. Many employers feel that the guidelines applied to testing are so rigorous, expensive, and time consuming that they have decided not to use this approach for hiring and promotion decisions.

The armed forces continue to use mental aptitude testing because of the presumed need to match an individual's mental capabilities with the demands of the job. The aptitude examinations are used to predict an individual's performance in the training program, which is a prerequisite for entrance into the related occupational specialty.[42] The services have estimated the aptitude requirements for the various formal training programs, but it is by no means clear that these estimates provide a good measure of what a job requires of an individual. Because

---

[41] See T. Anne Cleary et al., "Educational Uses of Tests for Disadvantaged Students," *The American Psychologist*, Vol. 30 (January 1975), pp. 15-40. This study reported that test scores have been misinterpreted from the beginning of the testing movement. Racial, regional, national, and sexual differences have been used without justification as data supporting genetic causation of differences in test scores and, in turn, as a basis for discrimination. This article covers relevant topics concerning uses of testing for educationally disadvantaged personnel such as (1) theory of human abilities, (2) test misuse and misinterpretation, (3) evaluation of "fairness" of tests in use, (4) alternatives to commonly used intellectual tests, and (5) tests of important qualities other than intelligence.

[42] It is felt that aptitude exams are most useful when applied to the evaluation of applicants for a training program, because of the capacity of the tests to measure the innate abilities involved in absorbing the training which the program provides. For more information, see Edwin E. Ghiselli, *The Validity of Occupational Aptitude Tests* (New York: Wiley, 1966).

of the many unmeasured, important factors involved, the services have found it difficult to relate the demands of the job to the aptitudes of individuals.

Consistent empirical evidence indicates that aptitude scores show only modest correlations with successful performances in training.[43] A classification decision based solely on mental aptitude examinations and moral and physical disqualifiers would appear incomplete. Additional measures of an individual's performance capability should be developed. For example, how important are the following to job performance: reliability, initiative, experience and training, and leadership potential?

The armed services are continuing to conduct research on personnel selection and are currently trying to determine the relevance of aptitude testing to matching the demands of jobs to the aptitudes of individuals. The services are also tackling the more generalized research problem of developing better predictors of successful job performance. More research is required in both areas. The results of the above research will have enormous implications for human resource management.[44]

Effects of Sending Personnel Classified as Not School Eligible to Formal School

Concerned by the imbalance of minority personnel across the rating structure, the Navy conducted a pilot program in 1973 and early 1974 that sent noneligible minority and nonminority recruits to Class A school. Assignment was based on the students' being motivated and having test scores no more than five points below the existing three-point waiver per test score

---

[43] John A. Sullivan, *Measured Mental Ability, Service School Achievement and Job Performance*, Center for Naval Analyses, Professional Paper No. 42 (Arlington, Va.: Center for Naval Analyses, 1970).

[44] For an analysis of traditional industry promotion criteria, together with two systems which are designed to match managers and supervisors with positions in firms, see Herbert R. Northrup, Ronald M. Cowin, et al., *The Objective Selection of Supervisors*, Manpower and Human Resources Studies, No. 8 (Philadelphia: Industrial Research Unit, The Wharton School, University of Pennsylvania, 1978).

Among the systems studied in this volume is the "assessment center method." Assessment centers seek to measure personal attributes such as creativity, human relations skills, behavior flexibility, leadership, need for approval of peers, decision-making ability, stress resistance, and other factors in order to select managers. A method similar to that of the assessment center method, which provides an *expanded* measure of an individual's capabilities and needs, should be designed by the military for enlisted personnel occupational selection.

component. Performance and progress of these students were monitored and compared with those of a random sample of eligible students.[45]

It was seen that the noneligible students had greater failure, setback, and disciplinary rates than their eligible counterparts had. Approximately 33 percent of the noneligible students were dropped from school, compared with 10 percent of eligible students. The majority of noneligible students did, however, graduate from A school.

Looking more closely at group differences, Bilinski, Standlee, and Saylor found that eligible students were more effective in performance of formal training school graduation requirements than were noneligibles. Nonminority noneligibles required more special help than minority noneligibles, and both subgroups required more help than eligibles. The researchers concluded that, with additional time in school and with remedial help, a majority of noneligibles, as previously defined, can graduate from school.

The above research has profound implications for distributing minority personnel more proportionately across all occupations. The research indicates that minorities' formal school performances are not reliably predicted by the traditional aptitude exams. Furthermore, so-called school noneligibles can be successfully trained in the more technical occupations. As already stated, additional research is needed to determine relevant eligibility criteria for formal school selection. Again, the question of inflated quality standards is raised. Does it really cost more in the long run to train an individual with lower aptitude when retention and the cyclical supply factors are taken into account? Even so, to what extent are the Navy and Marine Corps committed to upgrading minority personnel? What tradeoffs of cost are necessary to accomplish the stated goals for occupational distribution of minorities? To date, these questions remain unanswered.

ALTERNATIVE SELECTION AND TRAINING METHODOLOGIES

Since 1967, the military has expended considerable research effort attempting to increase the number of educationally dis-

---

[45] Chester R. Bilinski, Lloyd S. Standlee, and John C. Saylor, *Effects of Sending Minority Personnel Classified as Nonschool Eligible to "A" School*, Part 1, *"A" School Achievement* (San Diego, Calif.: Naval Personnel Research and Development Center, 1974), pp. 1-28.

advantaged personnel selected for technical training. Much of this effort has focused on the development of new test instruments that might be used in conjunction with, or in place of, the BTB. Generally, these instruments have been nonverbal, seemingly culture-fair, and nonacademically anchored in an attempt to avoid the influential role that education is suspected of playing in traditional aptitude testing.

Research is being conducted to develop ways in which to successfully predict potential job success without the use of traditional paper-and-pencil tests. It has been agreed that paper-and-pencil measures are neither adequate nor proper to demonstrate certain types of proficiency. The military is experimenting in this field of alternative job-potential measures. The Navy is conducting relevant research with respect to computerized tests and job-sample tests.

In some instances, low validities have been found for paper-and-pencil personnel selection tests against job performance criteria.[46] This may be the result of ability requirements for job performance which are too varied to be measured completely by written tests. Charles Cory has been evaluating computerized tests as predictors of on-the-job performance and has found that computerized tests provide accurate measurements of sequential information processing and memory skills, but offer no advantage over paper-and-pencil measures of perceptual speed.[47] The research is ongoing, but as of this date, computerized occupational selection examinations have not become policy.

Another possible alternative for the traditional mental aptitude test is the job-sample test, a practical examination composed of representative samples of the work involved in the job. A study of testing in education and industry notes: "Tests designed to assess specific abilities, such as typing or operation of a lathe, are important predictors of success on jobs. . . ."[48] Of all tests in the employment-selection and promotion field,

---

[46] Richard H. Lent, Herbert A. Auerbach, and Lowell S. Levin, "Predictors, Criteria and Significant Results," *Personnel Psychology*, Vol. 24 (1971), pp. 519-33.

[47] *An Evaluation of Computerized Tests as Predictors of Job Performance in Three Navy Ratings*, Vol. I, *Development of the Instruments* (San Diego, Calif.: Naval Personnel Research and Development Center, 1974), pp. 1-21.

[48] Milton G. Holmen and Richard Docter, *Educational and Psychological Testing* (New York: Russell Sage Foundation, 1972), p. 147.

these work-sample measures are generally considered by critics of testing to be least objectionable, probably because these measures appear to require the skills demanded in the given job.

The Navy has conducted research to determine whether demonstrated ability to learn selected aspects of a job can be employed as a predictor of ability to learn to perform the total job.[49] The machinist's mate rate was selected for initial research. A set of job-related miniature aptitude tests was constructed and administered to Navy recruits who had not scored high enough on the battery of selectors to be eligible for machinist's mate training in formal school. The recruits who performed satisfactorily on the job-related tests were placed on the job. Nine months later, their level of competence was compared with that of formal school graduates through work-sample performance test methods. The high aptitude machinist's mate students were found, on the whole, to perform significantly better on the performance criterion shipboard tests than did the low aptitude machinist's mates. In many individual cases, however, the low aptitude recruits performed as well as, or even better than, some of the formal school graduates on the performance criterion tests.

The results of interviews with the immediate work supervisors of the low aptitude sample indicated that 81 percent were performing at an acceptable level.[50] The next phase of this project involved administration of a second set of criterion tests to the initial sample after eighteen months of job experience. It was seen that, after eighteen months, there was no statistically significant difference between the predictive power of the miniature training and evaluative situations and the Navy tests. It was also seen that the mean scores of the low aptitude recruits improved more during the first and second follow-ups than did the performance of the A school control samples.[51] This was a fresh approach taken by the Navy to confront the problem of culture-fair testing. Similar research with similar tests, but with other Navy rates, will provide additional information.

---

[49] Siegel, Bergman, and Lambert, *Nonverbal and Culture Fair Performance Prediction Procedures*, pp. 19-54.

[50] *Ibid*, p. 55.

[51] Arthur I. Siegel and Wm. Rick Leahy, *Nonverbal and Culture Fair Performance Prediction Procedures*, Vol. III, *Cross Validation* (Wayne, Pa.: Applied Psychological Services, Inc., 1974), p. 37.

Currently, on a cyclical basis, the Navy conducts a task analysis of each occupational specialty which is supported by a formal school. Every three years, each formal school curriculum is reviewed and overhauled.[52] Questionnaires are distributed to selected commands in the fleet to determine the body of knowledge that is required for successful performance in a particular occupational specialty, as well as to determine in what respects the specialty has changed over the past three years. The returned questionnaires are used to validate the formal schools' training activities. A determination is made of whether modifications of the curriculum or teaching methodology are necessary for each of the formal schools. In this way, formal schools' curricula, teaching methodologies, etc., are modified on a regular basis to reflect the needs of the occupational specialties and the operational commands.

Self-Paced Training Methodology

Self-paced learning is a training methodology which allows less emphasis to be placed on mental aptitude testing. Variable course completion time frames are established for individuals enrolled in selected formal schools. The time allotted for a recruit to finish his formal schooling is a function of that individual's aptitude credentials. Self-paced learning allows individuals with different mental classifications to take part in the same formal training. Those with higher examination selector scores are expected to complete training more quickly than those individuals with lower selector scores. Reward, punishment, and remedial systems are set up to ensure the timely course completion by each individual according to that individual's expected time frame for course completion.[53]

The Navy is now using the self-paced learning method for many of its formal training courses. Formal school programs which have a large number of students and are of long duration are prime candidates for the transition to self-paced learning. Self-paced learning normally requires considerable student use of a computer system. Although this methodology allows

---

[52] Chief of Naval Technical Training, "Procedures for the Planning, Design, Development, Management of Navy Technical Courses" (Millington, Tenn.: Department of the Navy, 1974).

[53] Robert F. Lockman, *Enlisted Selection Strategies* (Arlington, Va.: Center for Naval Analyses, 1974), p. 47.

an individual to progress at his own pace, the average student is completing the necessary course work in one-third less time.

The self-paced learning methodology provides for a differential pace of course completion for students according to their aptitudes. Because of the differential time frame, however, less emphasis on mental aptitude is required. The implications of this system for minorities and nonminorities who could not qualify under the current aptitude requirements for such training are significant.

REMEDIAL EDUCATION

There is no question that adequate reading comprehension and mathematical ability are necessary for each recruit in order to embark on a successful military career. Research conducted in 1974 and 1975 at the Naval Personnel Research and Development Center demonstrated that a recruit's reading ability must be at approximately the seventh-grade level to complete boot camp successfully. As would be expected, higher reading and computational skills are necessary for formal training programs after boot camp.[54] The Navy enlisted advancement opportunity function indicates that good scores on the GCT and ARI, measuring reading and mathematical reasoning, respectively, are extremely important for advancement.[55]

The research results at the Naval Personnel Research and Development Center also indicated that approximately 7.5 percent of the total recruit population had a "reading grade level" below the sixth grade, and approximately 18.0 percent of the recruit population had reading skills below the eighth-grade level. The situation is far worse for minority personnel, however, than the above aggregated statistics indicate. Approximately 12.5 percent of the black recruit population have reading comprehension levels below the sixth-grade level. Almost 33 percent of the black

---

[54] Thomas M. Duffy, "Literacy Training in the Navy," in John D. Fletcher, Thomas M. Duffy, and Thomas E. Curran, *Historical Antecedents and Contemporary Trends in Literacy and Readability Research in the Navy* (San Diego, Calif.: Naval Personnel Research and Development Center, 1977), pp. 49-51.

[55] As part of a statistical analysis conducted to evaluate nonminorities' and minorities' advancement opportunities, a statistical model was developed to define the Navy enlisted advancement function. One of the conclusions drawn from the analysis is that individuals possessing capabilities which are measured by higher scores on the ARI and GCT exams have a distinct advantage in advancement opportunity. See chapter IV for further explanation.

recruit population do not have eighth-grade level academic credentials and, therefore, do not have the reading comprehension to complete formal training after boot camp successfully.[56]

From the above findings, it is obvious that minorities are severely restricted by the reading and computational requirements of formal schools. Without formal school training, individuals with poor educational backgrounds generally go to the soft-skill occupations. For these individuals, the services provide little job-skill upgrading.

Current Remedial Programs

Both the Navy and the Marine Corps currently have remedial education programs at all their recruit-training centers. Emphasis is placed on persons demonstrating deficiencies in reading and computational skills—below sixth-grade level.

The Marine Corps provides skill training in reading for those recruits falling below the 4.5 reading grade level. For the Navy, reading training is provided for all recruits having a tested reading grade level between 3.0 and 5.5. In fiscal 1975, an estimated 5.1 percent of the Navy recruit population had reading skills in this range. Recruits below a 3.0 reading level were considered poor candidates for short-term reading training and were recommended for discharge from the service. Recruits above a 5.5 reading level were considered by the Navy to have adequate reading skills.[57]

It is evident that two of the Navy's and Marine Corps' prime objectives in literacy are (1) to provide a level of literacy skill to all personnel so as to ensure fleet effectiveness and fleet safety and (2) to provide the literacy skills necessary for equal opportunity in attaining upward mobility and a successful career. Recent studies indicate a major gap between the reading requirements and the reading abilities in the Navy. Additionally, they suggest that literacy training to the 5.5 grade level is less than adequate for meeting the literacy demands in the service. A 5.5 grade level would seem neither to ensure fleet safety and effectiveness nor to provide the opportunity for upward mobility.[58]

---

[56] Duffy, "Literacy Training in the Navy," pp. 33, 40.

[57] *Ibid.*, pp. 51-59.

[58] Thomas M. Duffy and William A. Nugent, *Reading Skill Levels in the Navy* (San Diego, Calif.: Navy Personnel Research and Development Center, 1978).

Generally, the Navy's and Marine Corps' remedial programs have demonstrated only marginal success. The programs range from three to six weeks in duration. It is hard to imagine how a great deal of permanent success can be achieved when the remedial education programs are so short. Furthermore, the programs make no distinction between intellectual capability and educational inexperience in determining who gets remedial education benefits. Research by educational psychologists is needed to develop a good remedial education selection mechanism and a good remedial program in which lasting improvements in a recruit's reading and mathematical foundations can be achieved.

A prime requisite for any successful remedial reading program is a practical means of diagnosing the problem and its severity. At present, both the Navy and Marine Corps use indirect methods for reading problem diagnosis: (1) repeated difficulty on written examinations, such as the ASVAB and BTB, at the recruiting station and at boot camp; and (2) inability to perform satisfactorily at boot camp after reading an instruction manual or attending a lecture.[59] Both services might consider the institution of diagnostic tests to detect basic reading and quantitative skill deficiencies. The importance of the early detection of such individual deficiencies is paramount to an adequate system of human resource development.

The JOBS Program

The Navy and the Department of Health, Education and Welfare (HEW) announced in the spring of 1978 a new remedial training program for personnel not qualifying for Class A school. This joint research and development project, which is being partially funded by HEW, is entitled Job Oriented Basic Skills (JOBS).

Initially, this program will take recruits at the Naval Training Center in San Diego and provide them with special training in reading, writing, listening, computation, study behaviors, and other problem-solving tools. Additionally, these six-week courses will provide basic technical training in four general areas: ship propulsion, electronics, aviation mechanics, and administrative or clerical work. People who complete the JOBS program will be assigned to a Class A school.

---

[59] Duffy, "Literacy Training in the Navy."

Through this program, the Navy hopes to increase the number of young sailors who can be trained in technical skills.[60] After the first year, the Navy plans on enrolling students from the fleet into this program as well.

Is It Cost-Effective?

Whether it is cost-effective to upgrade the educationally disadvantaged must be noted. The armed services have, historically, sent to formal school training only those who scored highest on the mental aptitude exams. This method of formal school selection was designed for the high input/high turnover draft-motivated military. The argument for not providing a remedial education program is that the cost of upgrading individuals is an unnecessary expense additional to formal training. On the other hand, the all-volunteer force might very well require a different classification philosophy. It is true that the Navy and Marine Corps are currently filling their formal school quotas with qualified personnel without incurring the additional costs. But can this already qualified manpower pool be counted on if the economy expands? Furthermore, it is quite possible that higher aptitude individuals are no longer less costly overall, if indeed they ever were. Remedial education for those showing potential could prove cost-effective by virtue of increased productivity and retention. Research is needed to determine the validity of this hypothesis.

Those individuals selected for remedial education *must* have the intellectual capability and motivation to complete training. The cost of an effective remedial education program could be borne, to a large extent, by those persons receiving the education. In fact, it can be argued that this educational opportunity would be more fully appreciated and, therefore, successful if the recruit were made to pay for at least part of the education. Perhaps this could be accomplished by cuts in salary during the remedial education time period or longer or by an extension of the obligated service of those members participating in the program. Careful analysis must be conducted to determine what, if any, percentage of the training cost should be paid by the recipient and by what means.

Whatever remedial plan is devised, if additional funding is required, its implementation will rest with Congress. Recent congressional actions have been hostile to all military programs

---

[60] "Navy, HEW Launch Program to Improve Literacy Skills," *Navy Times*, June 12, 1978.

not directly related to national defense. Congressional cutbacks have reached into many military education programs such as General Educational Development (GED) testing, United States Armed Forces Institute (USAFI), and the Associate Degree Completion Program (ADCOP).[61] The military services are aware of the growing problem of basic educational deficiencies in today's recruit, and they correctly see this as an impediment to effective military force performance. Congress, as the ultimate supplier of funds, must be cognizant of this situation and its implications for effective manpower utilization and the ability of the military to achieve its mission. Furthermore, Congress must realize that the services are in direct competition with private industry in the labor market. Today's recruit is looking for training and the opportunity for advancement, and if he lacks the necessary basic skills, the services must help him to realize his objectives. If not, the traditional high input/high turnover rates of the military services will continue. Because minority personnel form a significant proportion of the service members who would be affected by remedial programs, the potential effect on their upgrading and mobility is substantial.

OFFICER CLASSIFICATION

Officers are normally assigned to a Navy warfare specialty or Marine Corps military occupation specialty at the time of commissioning. This designation is determined by the needs of the service and the officer's desires and qualifications. Some specialties, such as flight training and nuclear power training, require lengthy formal training programs, while others, such as Navy surface warfare, require relatively little preassignment training.[62]

Minority Officer Occupational Distribution

The occupational distribution of minority officers must be viewed in relation to the small numbers involved. Less than 1.8 percent of Navy officers are black, while approximately 3.6 percent of the Marine Corps officers are black. Table V-2 presents the absolute and relative breakdown for black officer occupational participation in the Navy and Marine Corps. In both the general officers and tactical operations officers categories, blacks are underrepre-

---

[61] The GED, USAFI, and other such programs were designed to further an individual's education while in the military.

[62] It takes an aviation trainee one year to eighteen months to earn his wings. Nuclear power training is approximately one year in length. Initial training for the surface warfare specialty is sixteen weeks.

TABLE V-2
Total Officer Personnel by Race and Occupational Group
in the Navy and Marine Corps
December 31, 1977

Department of Defense Occupational Group	Navy Total	Navy Black	Navy Percentage Black	Marine Corps Total	Marine Corps Black	Marine Corps Percentage Black
General Officers and Executive NEC	2,565	14	0.5	662	2	0.3
Tactical Operations Officers	12,121	175	1.4	11,677	306	2.6
Intelligence Officers	1,375	21	1.5	248	9	3.6
Engineering and Maintenance Officers	11,173	174	1.6	1,334	81	6.1
Scientists and Professionals	3,926	86	2.2	334	11	3.3
Medical Officers	2,889	44	1.5	a	a	a
Administrators	9,214	159	1.7	965	53	5.5
Supply Procurement and Allied Officers	3,387	75	2.2	1,301	104	8.0
Others	13,076	374	2.9	2,045	104	5.1

Sources: Bureau of Naval Personnel, MAPMIS 5350-0537-0A-01-01-68 (Washington, D.C.: Department of the Navy, January 27, 1978); U.S. Marine Corps, Headquarters, Manpower Planning, Programming and Budgeting Branch, August 1978.

[a] Serviced by the Navy in this occupational group.

sented. They are, however, substantially overrepresented in the occupational categories of scientists and professionals, supply procurement and allied officers, and "others."[63]

Table V-3 shows the distribution of black Navy and Marine Corps officers in selected professional occupations within the services. It is clear that blacks are not as well represented in these professional groupings as they are in some of the Department of Defense groupings of Table V-2.

The services recruit most of their doctors, dentists, lawyers, and chaplains from the civilian manpower pool.[64] Military doctors, dentists, lawyers, and chaplains enter the services either as practicing professionals or as students undergoing training pursuant to entry into the professions. The services do not provide in-house training for these professions. The dearth of minorities in the military who are in these professions, therefore, is a function of recruiting and of the small number of minorities in these professions.

The services train their own pilots. While on active duty, pilots serving in a flying status receive extra monthly pay. Those pilots choosing to leave the service after their initial obligation have traditionally found their skills highly marketable. The majority of civilian air transportation pilots are recruited directly from the armed forces.[65] As of October 31, 1976, the Navy had 162 black pilots, and the Marine Corps, 103. An Industrial Research Unit study of racial employment policies of the air transport industry found that, in 1969, only 0.3 percent of the pilots employed by scheduled airlines were black. At that time, 0.6 percent of the Air Force pilots and 0.1 percent of the Navy pilots (a total of 0.5 percent of the total armed forces pilots) were also black.[66] In this particular category, the Marine Corps and

---

[63] The "General Officers" category consists of Navy Admirals and Army, Air Force, and Marine Corps Generals. The "Executive NEC" category consists largely of senior line officers who are either in command or second in command of major shore installations and senior staff officers in selected high-level staff billets. The "other" category consists largely of officers who are students, unassigned and awaiting a billet opening, in special programs awaiting commissioning, or patients in hospitals.

[64] See Table III-6 for data concerning minorities in selected professions.

[65] Herbert R. Northrup, "The Negro in the Air Transport Industry," in Northrup et al., *Negro Employment in Land and Air Transport*, Studies of Negro Employment, Vol. V (Philadelphia: Industrial Research Unit, Wharton School of Finance and Commerce, University of Pennsylvania, 1971), Part Two, pp. 46-47.

[66] *Ibid.*, p. 46.

TABLE V-3

Selected Professional Occupations by Race
in the Navy and Marine Corps
September 30, 1976

Professional Occupation	Navy	Marine Corps
Aviation		
Total	18,624	6,402
Black	178	115
Percentage Black	0.9	1.8
Medical Doctors		
Total	3,628	a
Black	45	
Percentage Black	1.2	
Dentists		
Total	1,740	a
Black	17	
Percentage Black	1.0	
Lawyers		
Total	743	523
Black	11	18
Percentage Black	1.5	3.4
Chaplains		
Total	819	a
Black	12	
Percentage Black	1.5	

Sources: Column 2—Bureau of Naval Personnel (Pers-61), Report 5310-0371-Q61 (Washington, D.C.: Department of the Navy, September 30, 1976); column 3—U.S. Marine Corps, Headquarters (code MPH), "Marine Corps Quarterly Statistics" (Washington, D.C., March 31, 1978).

[a] Serviced by the Navy in these occupational groups.

the Air Force both exceed the record of the scheduled airlines. But the Navy falls below both other services and its civilian counterpart.

The Navy provides nuclear power training for the officers managing its nuclear reactor plants aboard ships and submarines. These officers have highly marketable skills, should they choose to leave the service. In addition, those who serve aboard submarines receive extra monthly pay. Officers in the nuclear power specialty also enjoy other special in-service benefits. First, they are eligible to receive a $20,000 bonus for agreeing to serve four years beyond their initial obligation. Second, they enjoy a faster promotion rate than officers in most other service occupations.

The entrance standards for aviation and nuclear power are more stringent than for the surface warfare occupational specialty. The nuclear power and pilot programs offer highly marketable skill training and additional pay incentives. Minority personnel are more poorly represented in these programs than in the surface warfare program.

Minority aviation officer candidates taking the Applicant Qualification Test (AQT) and Flight Aptitude Rating (FAR) test required for pilots experience the same problems that minorities typically face with respect to tests. Generally, they score lower on these examinations than their white counterparts. In addition, the biographical inventory section of the FAR attempts to measure how costly an aviation candidate's background matches the backgrounds of successful naval aviators. Since few naval aviators are minorities, it is quite possible that such a background matching process would not apply well to minority candidates.

The Navy and Marine Corps have substantially increased the percentage of black pilots in their forces during recent years. Interviews with Naval Flight Officer Program administrators, however, indicate that the efforts to integrate this program have not been fully successful. Minority participation has been characterized by a significantly higher attrition rate than has been experienced by nonminorities. Several reasons have been suggested for this phenomenon. They do not appear here in any priority ranking, nor is there any empirical evidence to quantify the frequency of the factors in cases of attrition. They do express the insights of the school staff into the problem.

As already stated, minorities do not score as well on the AQT sections measuring vocabulary and mathematical skills. Perhaps this indicates that minorities do not have as solid an educational background as nonminorities. The flight program curriculum is mathematically rigorous. Those persons not having a good mathematical foundation are sure to have difficulty.

Minorities tend to require more tutoring or extra instruction than nonminorities. Although instructors are generally willing to give extra help to any student who needs it, minorities are allegedly often hesitant to ask for it. Minority students may be intimidated by the fact that there are almost no minority instructors in the program, and this may impede the identification and resolution of minorities' problems. The psychological considerations in these situations are evident. Also, tutoring cannot be a continuous means of getting through the course material. Certainly, an individual may encounter particular concepts or problems which are hard to grasp and may seek extra help for their explanation, but the situation cannot be allowed to reach the point where every lesson requires out-of-class assistance.

Minorities have not adapted well to the programmed texts and self-teaching methods which the program emphasizes. The large quantity of reading and the need for extra tutoring can make keeping up with the course material extremely difficult. Once a candidate falls behind, it is particularly burdensome to catch up. The transition to the new learning environment of programmed texts is an entirely new experience for most. This is compounded by the sheer volume of necessary reading. Those persons who have reading deficiencies which never surfaced during their undergraduate studies are rudely awakened in this program.

Finally, many of the minority personnel have swimming difficulties. A large percentage have grown up in urban centers with socioeconomic backgrounds that precluded them from having the opportunity to learn to swim. Although virtually never the sole cause for withdrawal from the program, the additional psychological and physical burden it places on a student having to contend with a normally rigorous program can be the breaking factor.

The nuclear power training program is also highly selective. After careful prescreening, all potential candidates for the program are personally interviewed by Admiral H. G. Rickover, chief of the Bureau of Naval Reactors, who makes the final selections. Admiral Rickover selects largely from candidates with science and engineering undergraduate and graduate backgrounds. In recent years, he has made approximately one-half of his selections from among Naval Academy graduates.[67]

The nuclear power training program has two parts: nuclear power school and prototype training. Nuclear power school consists totally of classroom training to provide the student with both theoretical and practical knowledge of nuclear reactors and associated power plant equipment. During this six-month school,

---

[67] Bureau of Naval Personnel (Pers-402b), "Officer Personnel Retention Statistics," CNO/SECNAV Point Paper (Washington, D.C.: Department of the Navy, July 15, 1976).

TABLE V-4

U.S. Naval Academy Program Accessions

USNA Entering Class	Caucasian		Black		Other	
1972	1,260	93.0%	73	5.4%	16	1.2%
1973	1,224	89.0	112	8.2	36	3.0
1974	1,215	87.0	90	6.4	88	6.3
1975	1,245	89.0	55	4.0	96	7.0
1976	1,233	89.0	66	5.0	88	7.0

Source: United States Naval Academy Alumni Association, "Profile Class of 1980," *Shipmate*, Vol. 39, No. 8 (October 1976), p. 19.

students must pass regular examinations in all subjects, including mathematics, water chemistry, fluid dynamics, electrical engineering, and reactor physics. In addition, students must pass a comprehensive examination upon completing training. Those candidates who complete nuclear power school are sent to six months of prototype training during which they qualify as operators of a nuclear power plant similar to those aboard Navy ships. As discussed earlier, most program entrants have undergraduate or graduate backgrounds in engineering or the physical sciences, and over one-half are Naval Academy graduates. There are few minorities that meet these prerequisites.[68]

The Naval Academy graduates very few minority officers. Prior to 1973, few minorities were admitted to the Academy. Since 1973, the Academy has been somewhat more successful with minority accessions, but the total number is still small. Table V-4 shows the minority accessions for the years 1972 through 1976.

Given the small number of minorities with engineering degrees and the small number of minority Naval Academy graduates, the prospects for increased minority representation in the Navy's nuclear power branch appear dim in the near future. One course of action for the Navy to pursue in order to expedite a solution is to increase efforts to enlarge minority representation at the Naval Academy.

---

[68] See Stephen A. Schneider, *The Availability of Minorities and Women for Professional and Managerial Positions, 1970-1985*, Manpower and Human Resources Studies, No. 7 (Philadelphia: Industrial Research Unit, The Wharton School, University of Pennsylvania, 1977).

CHAPTER VI

Advancement

This chapter first discusses the relative standing of minority personnel in both the advancement system and the other areas of upgrading. Second, it examines the manner in which a service member progresses upward in the organization once he begins training in an occupational field.

ENLISTED MINORITY VS. NONMINORITY PERSONNEL PAYGRADE DISTRIBUTION

Table VI-1 shows the total number of personnel, the total number of blacks, and the percentage of blacks in each of the nine enlisted paygrades for both the Navy and Marine Corps at the end of 1977. Approximately 8.7 percent of the enlisted Navy and 17.6 percent of the enlisted Marine Corps are black. Obviously, blacks within these services are not proportionately distributed across paygrades, but are disproportionately concentrated in the lower grades.

In the Navy, 41 percent of the enlisted population are found in the non-petty officer rates (i.e., E-1, E-2, or E-3), and 59 percent are found in paygrades E-4 or below. Fifty-five percent of all enlisted blacks in the Navy, however, are found in the non-petty officer rates, and 72 percent of blacks are found in paygrades E-4 and below. Only 28 percent of blacks are found in paygrades E-5 through E-9, as compared with 41 percent of the total enlisted population. It is clear from the above statistics that blacks are disproportionately represented in the lowest paygrades and poorly represented in the higher paygrades.

The situation is not as severe in the Marine Corps, although inequities exist. Fifty-two percent of the total enlisted population, as compared with 60 percent of enlisted blacks, are in paygrades E-3 and below. Thirty percent of all enlisted personnel

TABLE VI-1
Total Enlisted Personnel by Race and Paygrade
in the Navy and Marine Corps
as of December 31, 1977

Paygrade	Navy Total	Navy Black	Navy Percentage Black	Paygrade	Marine Corps Total	Marine Corps Black	Marine Corps Percentage Black
E-9	3,447	147	4.2	E-9	1,234	127	10.2
E-8	8,373	431	5.1	E-8	3,323	456	13.7
E-7	30,443	1,768	5.8	E-7	8,552	1,156	13.5
E-6	64,052	3,326	5.2	E-6	14,016	2,033	14.5
E-5	80,872	5,367	6.6	E-5	25,125	4,097	16.3
E-4	80,574	6,855	8.5	E-4	29,488	4,382	14.9
E-3	96,202	10,294	10.7	E-3	39,200	7,053	17.9
E-2	52,214	6,425	12.3	E-2	28,977	5,899	20.4
E-1	42,847	5,503	12.8	E-1	22,195	5,081	22.9
TOTAL	459,024	40,116	8.7		172,119	30,284	17.6

Sources: Bureau of Naval Personnel (Pers-61), "Navy Wide Demographic Data Base for First Quarter FY-78 (1 Oct 77 to 31 Dec 77)," (Washington, D.C.: Department of the Navy, 1978); U.S. Marine Corps, Headquarters, Manpower Planning, Programming and Budgeting Branch, August 1978.

are in paygrades E-5 and above, but only 25 percent of blacks are in these paygrades. Although blacks are overrepresented in the lowest paygrades and underrepresented in the highest, the disparities are not as severe as those found in the enlisted Navy.

In the past, minorities have been promoted at a slower rate than nonminorities. Furthermore, as will be discussed shortly, there is room for improvement in the ability of the services' promotion functions to select the best qualified individuals for promotion. Thus, the Navy and Marine Corps face a difficult challenge in designing institutional mechanisms that will work to upgrade minorities, as well as provide a fairer promotion function.

PROMOTION PRACTICES AND STANDARDS

In the Navy and Marine Corps, the procedure for officer promotion and selection is essentially the same, but these services utilize different promotion criteria for their enlisted members. The Marine Corps has ceased to use written examinations as a promotion factor for E-6 and above. The Navy still relies heavily on advancement examinations for all enlisted personnel, although the weighting factor has been shifted downward.

Certain promotion criteria are common to both officer and enlisted advancement in the Navy and Marine Corps. First among these is a strict seniority system. Consideration for promotion, whether officer or enlisted, is based on minimum time periods of current rank. To be eligible for the next higher rank, an individual must have spent a specified period of time in his current rank.[1] For officers, there also exists a maximum number of times to be considered for promotion to the next rank. If an officer is passed over twice successively, then the length of the officer's career is automatically limited.

In order for an enlisted person to be advanced, he must be able to demonstrate certain skills as an indication of job competence. These skills are normally recognized and evaluated by the individual's supervisor or immediate superior and formalized through recommendation for advancement. Although such recommendations are normally based on objective factors, the potential for subjective bias does exist.

---
[1] Bureau of Naval Personnel, *Manual of Advancement*, BUPERS Instruction 1430.16A (Washington, D.C.: Department of the Navy, 1977), p. 3-1.

Advancement

Enlisted Advancement System Description

Within both services, promotions to paygrades E-2 and E-3 are essentially automatic upon completion of basic and intermediate training and time-in-grade requirements.[2] Promotion decisions for these paygrades are made at the local command level. Therefore, the commanding officer has the authority to promote individuals whom he deems qualified. Because the promotion decisions are not centrally administered, there are no quotas imposed by the Department of Defense on individual commands concerning the number of persons authorized to be promoted to paygrades E-2 or E-3. This is not the case, however, for decisions on promotions to paygrades E-4 through E-9. For these paygrades, the promotion decision mechanism is centrally administered. Ceilings are placed on the number of persons authorized for advancement. A strict quota system is imposed; therefore, competition can be, and often is, keen. Recommendations by the commanding officer for advancement are still required, but for the petty officer rates,[3] a recommendation for advancement is in itself but a prerequisite for further consideration.

The Navy and Marine Corps both agree that the selection criteria for promotion should be based on the "whole-man" concept, but each service emphasizes different factors in its promotion system. The factors considered by the Navy for advancement and their relative weights for paygrades E-4 through E-9 are shown in Table VI-2. The Marine Corps no longer uses a system by which specific weights are assigned. Figure VI-1 shows a sample worksheet used for developing the "composite score" used in determining promotion to E-4 and E-5. Promotion in the Marine Corps to E-6 and above is based on fitness reports (performance evaluations) and performance records and is determined by a closed selection board.

The Navy promotion system for paygrades E-4 through E-9 was designed to combine performance evaluation and merit ratings, scores on written tests, and seniority into a "final multiple score" to determine whom to advance. Individuals, as noted, must be recommended by their commanding officers in order to be considered for advancement. In addition, they must then pass

---

[2] The description of the services enlisted promotion system is taken, in large part, from the respective service manuals of advancement.

[3] Petty officer rates include all enlisted paygrades E-4 and above.

TABLE VI-2
*Weighting of
Navy Enlisted Advancement Factors for
Final Multiple Score Computation*

Factors	Paygrade				
	E-4 & E-5	E-6	E-7	E-8	E-9
Standard Score	80/35.0	80/30.0	80/60.0	80/50.0	80/40.0
Performance Evaluation	70/30.0	92/35.0	52/40.0	80/50.0	120/60.0
Length of Service	30/13.0	34/13.0	—	—	—
Service in Paygrade	30/13.0	34/13.0	—	—	—
Awards	10/4.5	12/4.5	—	—	—
Passed but Not Advanced (PNA) Points	10/4.5	12/4.5	—	—	—
TOTAL	230/100.0	264/100.0	132/100.0	160/100.0	200/100.0

Source: Bureau of Naval Personnel, *Manual of Advancement*, BUPERS Instruction 1430.16A (Washington, D.C.: Department of the Navy, 1977), p. 1-6.
Note: Read above chart as: Maximum allowable "final multiple score" points/percentage of total "final multiple score."

the advancement examination before being further considered. Once an individual is recommended for promotion to paygrades E-4, E-5, or E-6 and passes the examination, the selection decision is based on the weighting procedures for the six factors listed in Table VI-2. Those with the highest final multiple scores are advanced. For advancement to paygrades E-7, E-8, and E-9, the final multiple score is based on the standard score and performance evaluations and is used to determine selection board eligibility only. Final selections are made by the selection boards. Those candidates who pass the exam but, because of reduced promotion quotas, are not advanced receive passed-but-not-advanced (PNA) points, which are applied to the next competition.

As stated, there are distinct differences between the Navy and Marine Corps, both in criteria used for promotion selection and in the relative emphasis given to the promotion factors. First, the Marine Corps does not rely on paper-and-pencil advancement exams in the same manner as the Navy. A series of exams

Advancement

FIGURE VI-1
*Composite Score Computation for
Marine Corps Promotion to
Corporal (E-4) and Sergeant (E-5)*

Line No.			Rating
1.	Rifle marksmanship score: _____		_____
	(+)		
2.	PFT: _____ minus _____ = (−) _____	(Difference)	_____
	(Score) (Passing)		
3.	Essential subjects: (Number passed = _____)		_____
4.	Subtotal		_____
5.	GMP score (line 4 divided by _____)		_____
6.	GMP score (from line 5)	_____ X 100	_____
7.	Average Duty Proficiency	_____ X 100	_____
8.	Average Conduct	_____ X 100	_____
9.	Time in Grade (months)	_____ X 5	_____
10.	Time in Service (months)	_____ X 2	_____
11.	DI/Recruiter/MSG Bonus (100 points)	_____ X 1	_____
12.	Self-Education Bonus	_____ X 10	_____
13.	Composite Score (Sum of lines 6 thru 12)		_____

Source: U.S. Marine Corps, Headquarters, *Marine Corps Promotion Manual* (Washington, D.C., 1977).

known as "essential subjects" are given periodically to personnel E-7 and below. The essential subjects are a means to ensure maintenance of infantry proficiency. They are also a factor in the advancement function for advancement to E-4 and E-5; however, this factor does not count nearly as heavily as in the Navy. In fact, one could conceivably have a zero and yet have a composite score suitable for advancement. As is evident from Figure VI-1, average duty proficiency and average conduct are both important in determination of the composite score. As was seen, the Navy places less emphasis on these latter two areas.

For promotion decisions concerning the advancement of senior enlisted personnel, selection boards are used in the Navy for promotion to paygrades E-7 through E-9 and in the Marine Corps for promotion to E-6 through E-9. The selection boards of each service reside in Washington. In the Navy, the board reviews performance evaluations, special abilities, achievements, and test scores. In the Marine Corps, major emphasis is given to fitness reports and performance records. The selection boards in each case meet for a designated period of time and review each individual record. Candidates for the next paygrade are then selected according to their relative standing and the Department of Defense promotion quota.

The Role of Performance Evaluations

Performance evaluations play an extremely important role in the promotion process. Performance on the job, to a large extent, determines whether an individual is recommended for promotion, and a recommendation for promotion is a prerequisite for further promotion consideration. Furthermore, these evaluations determine an individual's reenlistment eligibility.

It is through these subjective evaluations that direct bias can affect an individual's ability to compete with his peers for advancement. Although these reports are written by an individual's supervisor, who in turn passes the report up the chain of command for approval before it is finally sent to headquarters in Washington, the procedure is certainly not without weaknesses. Indeed, a key factor in any system of evaluation is that the individual being evaluated knows how the system operates, its significance for his advancement, and, most important, the contents of the periodic evaluation report. In this manner, the individual is made aware of his performance for the preceding period and, it is hoped, will receive deserved praise or constructive criticism. In the Navy, this process is accomplished by requiring the evaluated individual's signature on the report itself and a face-to-face consultation between the individual and his immediate superior to review and discuss the evaluation marks.

The Marine Corps diverges markedly from the Navy in this practice. The Marine Corps maintains an official policy of not showing marines (officer *or* enlisted) their fitness reports unless they are derogatory, in which case the individual must sign the report. According to the Marine Corps, the arguments for this procedure are twofold.[4] First, if the evaluator knows

---

[4] Marine Corps' arguments for not showing or discussing an individual's performance evaluation with him are taken from interviews with Marine Corps officers conducted at U.S. Marine Corps, Headquarters, Washington, D.C., June 1976.

Advancement

he must show the report to the individual, he will tend to inflate the scores. Second, if a marginal performer knows his superior considers him marginal, then his performance, at least under battlefield conditions, may prove more unreliable.

Unfortunately, because face-to-face sessions are required only when there is a derogatory evaluation, the Marine Corps system can lead quite easily to a number of evaluations which damn with faint praise, but are not sufficiently derogatory to require a face-to-face session. The wording of descriptive paragraphs within the evaluation report is so crucial that the insertion of a "but" or "however" or other qualifiers will almost categorically remove an individual from further consideration for advancement. Such phrasing may not even be considered derogatory by the evaluator.

Subtle racial discrimination can certainly exist in performance rating and work to the detriment of minorities. The need for good race relations is obvious. The effectiveness of the services' human relations training is directly tested in this aspect of the promotion decision.

The Role of Advancement Exams

The Navy has designed separate advancement examinations for each of over eighty occupational specialties and for each paygrade within that specialty. These technical examinations ostensibly measure an individual's theoretical and applied knowledge of his occupation. The percentages of applied and theoretical questions found on the tests vary with each specialty. For example, advancement exams in electronics test heavily for theoretical knowledge, whereas the exams given to cooks test heavily for applied knowledge. All of these written examinations, however, should successfully measure an individual's potential to perform at the next higher paygrade within his occupation.

Navy-wide examinations are given twice yearly for paygrades E-4, E-5, and E-6 in March and September; once yearly for paygrade E-7 in January; and once yearly for paygrades E-8 and E-9 in November. The contents of the examinations are as follows for the designated paygrades:

1. Petty officer third class and petty officer second class (E-4 and E-5)

(a) Military/Leadership examination—100 questions

(b) Professional examination—150 questions

2. Petty officer first class and chief petty officer (E-6 and E-7)

    (a) Professional examination—150 questions

3. Senior chief petty officer (E-8)

    (a) Examination consists of 50 technical, 20 military and collateral duty, 20 special aptitude, 20 supervision, 20 communication (verbal reasoning), and 20 problem-solving questions.

4. Master chief petty officer (E-9)

    (a) Examination consists of 45 technical, 30 military and collateral duty, 35 administration, 20 communication (verbal reasoning), and 20 problem-solving questions.[5]

The Navy advancement system was designed to be a combination of the best features of three basic types of promotion systems to provide credit for knowledge, performance, and seniority. The advancement exams have, however, demonstrated in the past an influence upon the promotion decision that exceeds their official policy weighting.[6]

ADVANCEMENT EXAM DOMINATION OF PROMOTION SELECTION

As shown in Table VI-2, the components of the official Navy enlisted advancement system are technical knowledge (advancement) examination, 40 percent; on-the-job performance marks, 25 percent; time in grade, 10 percent; time in Navy, 10 percent; medals and awards, 7.5 percent; and bonus points for exam performance in previous advancement competition in which the candidate passed the exam, but, due to reduced promotion quotas, was not advanced (PNA), 7.5 percent.

Research began in 1972 to determine to what extent the existing operational weighting procedures meet the requirements

---

[5] Bureau of Naval Personnel, BUPERS Instruction 1430.16A, pp. 1-5, 1-6.

[6] David W. Robertson, "Improving Equal Opportunity in Advancement for Minority Groups" (Paper presented at the 35th Military Operations Research Symposium, U.S. Naval Academy, Annapolis, Maryland, July 1-3, 1975), pp. 6-7.

stated in the Navy's *Manual of Advancement*.[7] The results of the research revealed that, prior to February 1974, the factor-weighting procedures did not effectively provide credit according to the above published policy. Rather, the actual effect was an examination weight in excess of the 40 percent policy weight. For example, the examination was found to carry all the weight in 61 percent of the occupational and paygrade groups competing in August 1970, 51 percent of the groups in February 1972, and 35 percent of the groups in February 1973.

The excessive weighting of the examination factor was attributable to high examination cut-off scores. The cut-off scores were set to eliminate from further promotion competition those who were not considered qualified by virtue of their professional examination performance. In many cases, however, the excessively high cut-off scores failed the majority of candidates and, therefore, singularly defined the advancement function. In many cases, then, the examination results alone served to determine who would be advanced and who would not. Thus, the official weighting procedures were ineffective.

As a group, minorities do not perform as well on the enlisted advancement exams as nonminorities do.[8] The fact that the high exam cut-off scores frequently eliminated from further consideration a majority of candidates, especially in the higher paygrade competitions, served to bias the procedure against minority groups. That is, the higher the cut-off score, the greater the disproportionate percentage of minorities disqualified from further consideration. In the February 1972 competitions, the proportion of blacks failed at paygrade E-4 was twice that of whites: 38 percent and 19 percent, respectively. At paygrade E-6, 71 percent of the blacks were failed, compared with 55 percent of nonminorities, and black advancement was approximately one-half that of nonminority advancement.[9]

The Navy's argument for requiring cut-off scores is that it does not want to promote individuals who do not have an adequate technical background in their occupational specialty. Yet,

---

[7] David W. Robertson, Jim James, and Marjorie H. Royle, *Comparison of Alternative Criteria and Weighting Methods for the Enlisted Advancement System* (San Diego, Calif.: Navy Personnel Research and Development Center, 1972); Robertson, "Improving Equal Opportunity in Advancement."

[8] Christopher Jehn and Marcella Wojdylak, *Blacks in the Navy: Some Background Information* (Arlington, Va.: Center for Naval Analyses, 1973), p. 6.

[9] Robertson, "Improving Equal Opportunity in Advancement," p. 7.

the cut-off scores have been determined somewhat arbitrarily in order to fill quotas and, therefore, have fluctuated as a function of the manpower pool. There has been no research conducted to indicate at what point the cut-off score should be set. If the Navy insists on using cut-off scores, it would appear that they should be validated. This would require that research be conducted to provide reliable evidence that those not scoring above the designated cut-off score are incapable of performing at the next higher paygrade satisfactorily.

Setting excessively high cut-off scores has served to cripple the effect of the other factors by removing from competition those individuals who most need to be measured by the other policy factors. Unfortunately, minorities are disproportionately represented in this category.

If it is assumed, for the time being, that the Navy's factor-weighting policy is the best selection device, then the imprecision of the above mentioned cut-off score weighting system could have failed to select the best all-around qualified candidates. This statement holds true for nonminority and minority candidates. With the advancement examinations serving as the sole criterion, it is possible for the poorest on-the-job performers to be advanced rather than the best or even average performers. In fact, it was found that, during the 1972 advancement competitions, the job performance evaluations of the blacks not advanced were slightly higher than both the black and white advanced groups.[10]

In February 1974, the chief of naval personnel directed the Naval Personnel Research and Development Center (NPRDC) to design the advancement function so that the published factor-weighting procedures would be adhered to. This was done, and as of August 1974, the advancement function weights are as designated in the first paragraph of this section.

By lowering the exam cut-off scores, the advancement system is now found to work in accordance with policy. The system combines an individual's performance, experience, and technical knowledge to select for advancement those who best meet the whole-man criteria. Furthermore, improvement in the ability of the promotion system to achieve policy weights is found to increase substantially the promotion opportunity of black personnel.[11] That is, a more proportionate percentage of minorities now pass

---

[10] Significant at the 0.001 level of significance.

[11] Robertson, "Improving Equal Opportunity in Advancement," pp. 8, 9.

the advancement examination and are, therefore, further considered for promotion. Nevertheless, it is still possible for the poorer performers to be advanced rather than the top performers in the Navy system because most of the factor weighting is placed on the advancement exam. Furthermore, the exam cut-off scores, although lower, still disqualify a disproportionate percentage of minorities.

ALTERNATIVE TESTING METHODS

The armed services have conducted, and continue to conduct, a considerable deal of research on personnel selection. The ability of advancement examinations to reliably predict performance at the next higher paygrade is being carefully evaluated. Much effort has focused on the development of new test instruments that might be used in place of the current examinations. Similar-item-difficulty tests and job-sample tests are two methods of personnel selection that are being studied by the Navy.

Similar-Item-Difficulty Tests

In 1975, an initial investigation of the feasibility of constructing advancement exams which are assured of being unbiased was published.[12] Specifically, the possibility of constructing exams containing only questions which are similar in difficulty for blacks and nonminorities was studied. For each type of advancement test, the proportion of questions found to be similar in difficulty for blacks and nonminorities varied from about one-half to six-sevenths of the 150 questions. These questions were generally concentrated in the difficult range and were applied or factual in content, not theoretical or conceptual. Tests constructed of these questions were found to reduce, but not to eliminate, differences in exam scores between blacks and whites.

The development of advancement exams with questions similar in difficulty is not recommended; the concentration of similar-difficulty questions in the difficult or guessing range will downgrade test quality. It is also quite possible that questions largely limited to factual content may not cover all of the conceptual knowledge deemed necessary for advancement in a particular occupational specialty.

---

[12] David W. Robertson and Marjorie H. Royle, *Comparative Racial Analysis of Enlisted Advancement Exams: Item-Difficulty* (San Diego, Calif.: Naval Personnel Research and Development Center, 1975).

Job-Sample Tests

Research is also being conducted to develop ways in which to successfully predict potential job success without the use of traditional paper-and-pencil tests. Generally, this research has centered on nonverbal, seemingly culture-fair, and nonacademically based testing methods in an attempt to reduce the influential role that education is suspected of playing in advancement examinations. It is felt by many that paper-and-pencil measures are neither adequate nor proper to demonstrate certain types of proficiency.[13] Furthermore, the emphasis on job relatedness in court decisions and federal test regulations has increased interest in alternative measures of job potential.

Performance or job-sample tests are being investigated in the private sector and in the academic community as possible alternatives to traditional testing.[14] Job-sample tests are practical exams composed of representative samples of the work involved in the job. As stated in chapter V, the Navy is also conducting research on such tests [15] to determine whether demonstrated ability to learn selected aspects of a job can be employed as a predictor of ability to learn to perform the total job.

How well do job-sample tests predict performance on the job? This question has not been answered conclusively. Research is ongoing. The Navy's initial results indicate, however, that job-sample testing can be a useful factor in selecting individuals for promotion, especially for manually oriented occupations. The validity of this method of measurement for occupations requiring abstract reasoning has yet to be determined. It is quite possible, though, that job-sample tests will be found more reliable in their prediction of on-the-job performance than the traditional paper-and-pencil exams for some occupations. This remains to be seen. Initial evidence, however, indicates that, when this form of measurement is used in the selection process, the advancement potential of minorities is increased.

---

[13] J. E. Campion, "Work Sampling for Personnel Selection," *Journal of Applied Psychology*, Vol. 56 (1972), pp. 40-44; Richard H. Lent, Herbert A. Auerbach, and Lowell S. Levin, "Predictors, Criteria and Significant Results," *Personnel Psychology*, Vol. 24 (1971), pp. 519-33.

[14] Campion, "Work Sampling," p. 44.

[15] Arthur I. Siegel, Brian A. Bergman, and Joseph Lambert, *Nonverbal and Culture Fair Performance Prediction Procedures*, Vol. II, *Initial Validation* (Wayne, Pa.: Applied Psychological Services, Inc., 1973), pp. 19-54.

ALTERNATIVE WEIGHTING OF PROMOTION SYSTEM FACTORS

The Navy's rationale for heavily weighting the written advancement examinations stems from the belief that these exams are the best predictor of performance at a higher paygrade. Its argument is twofold.[16] First, the examination is composed of questions derived from the specific skills required for advancement in a particular occupation; second, the examination is a *recent* measure, as opposed to the performance evaluation average which can extend back over a much larger period of time.

On the other hand, some recent research challenges the hypothesis that the advancement exam is the most reliable predictor of who will perform best at the next higher paygrade. The results of the Naval Examining Center's validation study indicate that the on-the-job performance evaluation may be a better selector than the advancement exam.[17] If the results of this research are found to be the general rule, then the Navy's rationale for heavily weighting the exam is contradicted. A good case can then be made for increasing the weight of the performance variable in relation to the weight of the examination variable. Furthermore, in some cases, the best performers are not the best test takers. Those who have not performed as well as others in the past might be advanced, while those who have demonstrated strong on-the-job performance might not.

All persons taking the advancement examination are recommended for promotion by their command, but because of the present factor-weighting system, an individual's technical expertise and test-taking ability are rewarded more heavily than the commanding officer's appraisal of the individual's on-the-job performance. The commanding officer's appraisal can be biased, but so perhaps can be the tests. If it is true that on-the-job performance evaluation is the best predictor of performance at the next higher paygrade, then a strong argument can be made for placing additional emphasis on the job evaluation variables.

The full potential of the Navy advancement system to upgrade minorities cannot be fully realized as long as the advancement examinations are weighted as heavily as they are today. This is

---

[16] Argument made by Naval Examining Center (NAVEXAMCEN) personnel. The Naval Examining Center is responsible for design and implementation of personnel testing.

[17] NAVEXAMCEN study conducted in 1970.

the case because minorities, as a group, score significantly lower on these exams than nonminorities score. On the other hand, the performance model results, discussed in chapter IV, indicated that there is little difference between race-ethnic groups' scores on the current performance evaluations. Therefore, if the weight of the on-the-job evaluations is increased and the weight of the advancement examinations decreased (but not dropped entirely), minorities' upgrading opportunities are substantially increased.

MILITARY JUSTICE

The armed services' judicial systems derive their authority from the Uniform Code of Military Justice (UCMJ). The UCMJ was passed by Congress in 1950 and was significantly amended in 1968.[18] There are four levels under the UCMJ, classified by their punishment limits. The first three levels are the General, Special, and Summary Courts-Martial. These three judicial proceedings are convened for serious offenses. Those found guilty under any of the above three proceedings are severely punished and may even be discharged under conditions which are less than honorable. The fourth level of punishment is imposed nonjudicially by a unit commanding officer or officer-in-charge under Article 15 of the UCMJ. This fourth level is referred to as nonjudicial punishment (NJP). NJP actions are usually taken for minor offenses. The authorized punishment level varies with the rank of the commanding officer or officer-in-charge. The maximum allowable punishment under NJP is correctional custody for thirty days, forfeiture of one-half of one month's pay per month for two months, and reduction to E-1 for enlisted personnel in paygrades E-4 and below or reductions of one paygrade for enlisted personnel above paygrade E-4.

Enlisted personnel are disciplined far more frequently under NJP than under other punitive levels. Furthermore, although black sailors and marines are more likely than their white counterparts to be punished at all four punitive levels, it is NJP that is most frequently the target of racial discrimination charges. This is partially because of the operating procedures of the NJP system. Commanding officers, for example, are granted wide discretion in bringing charges and dispensing punishment under

---

[18] See Office of the Assistant Secretary of Defense, *Report of the Task Force on the Administration of Military Justice in the Armed Forces*, Vol. I (Washington, D.C.: Department of Defense, 1972), p. 7.

Article 15 of the UCMJ. For these reasons, this investigation of military justice will concern itself primarily with NJP.

The statistical analysis results discussed in chapter IV indicate that an individual's disciplinary record strongly influences his promotion opportunities. Unfortunately, a disproportionate share of minorities, relative to nonminorities, are reduced in paygrade during the course of their enlistment. Because disciplinary action determines, to a large extent, whether or not an individual is advanced, the impact of a disproportionate number of "busts" on minorities' advancement is severe.

Race-Ethnic Groups and Military Justice Offenses

A comparative study of civilian and military criminal justice systems revealed that, in the civilian system, blacks are more likely than whites to be arrested, convicted of crimes, and incarcerated. A similar situation, not surprisingly, is found to exist in the military. The proportion of blacks in military correctional facilities exceeds their proportion within the services, but the proportion of blacks in civilian correctional facilities is higher than that in military facilities.[19]

An earlier study conducted by the Department of Defense foretold the above results.[20] A disproportionate number of black servicemen were found to be involved in military justice actions. According to the study, blacks are more likely than whites to be charged with confrontation or status offenses.[21] Whites are more likely to be charged with drug-related and other military and civilian offenses.[22] These findings were confirmed for Navy personnel specifically in a study conducted by the NPRDC.[23]

---

[19] John J. Coursey, Johnnie Daniel, and R. J. Landman, Sr., *Analysis of the Military and Civilian Criminal Justice System and Final Report* (Washington, D.C.: A. L. Nellum Associates, Inc., 1973).

[20] Office of the Assistant Secretary of Defense, *Administration of Military Justice in the Armed Forces*, Vol. I, pp. 25-30.

[21] Such as failure to obey an order or insubordinate conduct toward a noncommissioned officer.

[22] Such as misbehavior of sentinel or larceny.

[23] Patricia J. Thomas, Edmund D. Thomas, and Samuel W. Ward, *Perceptions of Discrimination in Non-judicial Punishment* (San Diego, Calif.: Naval Personnel Research and Development Center, 1974).

Causes of Disparate Offense Rates

The causes of the disparate offense rates between race-ethnic groups are not clear. Are these differences due to either institutional or interpersonal racial discrimination? As noted earlier, the nonjudicial punishment system is a common target of such charges. At present, however, there is no evidence to support these charges.

Interviews with personnel at all levels of the Navy and Marine Corps indicated considerable confidence in the fairness of unit commanding officers in handing out punishment for individuals of different races. Furthermore, research conducted by the NPRDC found no differences in disciplinary actions across race-ethnic groups for similar offenses.[24] Although these findings clearly do not prove equality of treatment, they do suggest that, once an offense is officially processed, it is handled by the commanding officer without regard to race. It is, however, possible that, in cases in which offenses are not officially processed for NJP, minorities are more often and more severely disciplined than nonminorities, but there is no evidence to support this hypothesis.

Race-Ethnic Groups, NJP Rates, and Organizational Effectiveness

A recent study was conducted to correlate NJP rates aboard Navy ships with certain aspects of organizational effectiveness.[25] All of the aspects, or variables, were found to be correlated negatively with NJP rates so that the higher the rating given a ship's command for organizational effectiveness, the lower the NJP rate tended to be. The two factors found to be most significantly related to NJP rates were "supervisory support" and "communications flow."

Two independent research studies revealed a significant difference between blacks' and nonminorities' perceptions of their command's communications flow and supervisory support.[26] Nonmi-

---

[24] *Ibid.*, p. 11.

[25] Kenneth S. Crawford and Edmund D. Thomas, *Human Resource Management and Nonjudicial Punishment Rates on Navy Ships* (San Diego, Calif.: Naval Personnel Research and Development Center, 1975).

[26] Warrington S. Parker, Jr., *Differences in Organizational Practices and Preferences in the Navy by Race* (Ann Arbor, Mich.: Institute for Social Research, University of Michigan, 1974); Thomas et al., *Perceptions of Discrimination*.

norities were found to view their command's organizational effectiveness (in terms of the above two factors) more favorably than do minorities. Furthermore, those who reported having an immediate supervisor of the same race were found to have more favorable perceptions of their command's organizational effectiveness than those who reported having an immediate supervisor of a different race.

Although providing no conclusive answers to the disparate NJP rate question, the above discussion does provide some insight into the problem. NJP rates are strongly related to how well the personnel thinks the command manages human resources. Communications flow and supervisory support are the two elements of organizational effectiveness (human resource management) most closely related to NJP rates. With respect to these factors, blacks perceive their unit's environment less favorably than nonminorities perceive it.

The implication is that a serious problem appears to exist between enlisted blacks and their immediate supervisors, particularly when the supervisors are white. It appears that blacks feel that they are not being supported by their supervisors, and that they are being left out of the formal communications channels; it appears that the friction at this supervisor/subordinate level may be leading to excessive black NJP rates. This assessment seems particularly plausible when it is recalled that blacks are most likely to be reported for confrontation or status offenses, such as failure to obey an order or insubordinate conduct toward a noncommissioned officer. This attitude, whether or not based upon evidence, could thus be a negative promotion factor.

LATERAL TRANSFER

The Navy has designed a personnel management system (Career Reenlistment Objectives, CREO) to match manpower requirements and advancement opportunity with occupational categories. Each of the Navy occupational specialties is now assigned to one of the following three categories: open rating, neutral rating, and closed rating.[27] The open rating category (fast promotion rate occupations) includes those occupations which pro-

---

[27] Navy ratings (occupations) are broken down into Career Reenlistment Objectives (CREO) categories in Bureau of Naval Personnel, "Career Reenlistment Objectives (CREO)," BUPERS Instruction 1133.25C (Washington, D.C.: Department of the Navy, 1975).

vide the greatest amount of advancement opportunity because of manpower shortages in paygrades E-4 and above. The closed rating category (slow promotion rate occupations) includes those occupations which provide the least amount of advancement opportunity because of overmanning. In the statistical analysis conducted to evaluate nonminorities' and minorities' advancement opportunities, occupational classification was found to be significant for an individual's promotion rate.

Blacks are severely underrepresented in the occupations with the greatest advancement potential. In the sample, 42 percent of all personnel are assigned to an open rating, but only 27 percent of the blacks in the sample are so assigned. Minorities are therefore not proportionately distributed across occupations nor as fortunate in advancement opportunity as nonminorities are.

One possible means through which minorities can be redistributed is the services' lateral transfer programs. According to the Bureau of Naval Personnel, "a change in rating is a lateral change in occupational skill without change in paygrade."[28] The Navy and Marine Corps have similar policies for lateral specialty changes between reenlistments. The Navy's policies will be discussed here for illustration.

To be eligible for a change in occupational specialty, one must request such a change and meet the following criteria:

1. The requester must be a petty officer first class (PO1— E-6) or below. In the case of PO1s, the needs of the service must be great in order to grant the change in rating since the person possesses valuable skill and experience in his present occupation.

2. The requester must have less than 15 years' active service.

3. The new occupational specialty must be on the open rating list. Recall that the open rating category includes those occupations that have a shortage of personnel.

4. The requester shall be eligible to receive clearances if necessary.

5. If training is required for entrance into the new specialty, the requester must meet the minimum BTB or ASVAB score requirements.

---

[28] Bureau of Naval Personnel, *Bureau of Naval Personnel Manual*, NAVPERS 15791B (Washington, D.C.: Department of the Navy, 1969), Article 2230180.

Advancement

6. If the new specialty requires, for any reason, that the requester be transferred from his present command, he must be eligible for transfer.
7. The requester must not be receiving any form of bonus pay in his present rating.
8. The requester must be recommended by his commanding officer.[29]

If a person satisfies all of the above criteria, he may have his occupational specialty changed by one of the following methods:

1. By command administrative action.
    (a) Individual commands may authorize change of apprenticeships if there is greater need in the new specialty, there is a valid vacant billet in the command, and the requester is fully qualified.
    (b) Training commands may authorize occupational specialty changes upon successful completion of training.
2. By "in-service training." This method involves on-the-job training and outside study by the individual. Lateral transfers are authorized only from an overmanned specialty to an undermanned specialty.
3. Through completion of formal school training.
4. By successful completion of a Navy advancement exam. If a member in an overmanned occupation can pass the advancement exam in an undermanned occupation at the same paygrade level and is otherwise qualified, a change of specialty may be authorized.[30]

From the above criteria and methods, it is obvious that opportunities to transfer laterally into a different occupational specialty are severely limited for career Navy personnel. Still, personal diligence and patience may be rewarded if the desired change of occupation is granted.

The lateral transfer programs are not currently being used specifically to redistribute minorities from the overmanned to the undermanned occupations. The primary constraint to any such action is the test score qualifying requirement. Generally, minorities in the overmanned occupations do not qualify under the test

---

[29] *Ibid.*

[30] *Ibid.*

cut-off score requirement for lateral transfer to the undermanned technical occupations.

MANPOWER CONTROL VS. AFFIRMATIVE ACTION PROMOTION PROGRAMS

The military services are constrained in their manpower-planning policies by congressional action which yearly determines the budget of each of the services. The recent reduction of all the armed forces from their peak Vietnam-era strength to peacetime levels has significantly affected the services' ability to maintain promotion levels and has created other manpower control problems as well.

Both the Navy and the Marine Corps have programs under way which are aimed at manpower control planning. The Navy Enlisted Occupational Classification System (NEOCS) and the CREO system are helping to normalize promotion training. Often in the past, advancement in particular occupational fields has stagnated. When turnover in the upper ranks slows or ceases, promotions are less frequent. Naturally, the competition for promotion becomes very intense, and many who do not succeed leave the service in frustration. The Navy recognizes this loss of good manpower and has implemented NEOCS and CREO in an effort to match manpower requirements and advancement potential more closely with occupational categories.

The recruitment and retention of the most qualified minorities has proved to be a major problem for the services. Because of the all-volunteer military, the services have had to compete with private industry for the best nonminority and minority personnel talent. Many officers and enlisted men (both minority and nonminority) sincerely feel that the services cannot provide the recognition and advancement opportunities available in the civilian labor market. The affirmative action promotion policies of private industry have directly contributed to this feeling among minorities. The strongest pull away from the military is certainly being felt by minority officers, whose experience and education can be much more quickly developed and rewarded by aggressive private sector firms than by the military. The armed forces have also found it difficult to retain the most qualified minority enlisted men.

Pursuing affirmative action promotion policies involves identifying competent minorities and placing them in varying levels

Advancement

of supervisory and managerial positions as fast as their skills develop. The objective, of course, of such policies is to increase the participation rate of minorities at all levels of the organization. Needless to say, the institution of such programs has not been easy or without negative reactions.

The armed services do not have a program which specifically places special emphasis on the rapid development and early promotion of minorities. An aggressive affirmative action promotion program would provide the services with the ability to distribute minorities proportionately across all paygrades much more quickly than is currently possible. It is safe to assume, however, that nonminorities would react negatively to a modification of the seniority-based promotion system which only benefited minorities, and politically, it is probably not feasible, even if it were practical.

EDUCATIONAL OPPORTUNITIES AND CAREER COUNSELING

The preceding sections of this chapter have been devoted to the exploration of institutional factors which play influential roles in the upgrading of first-term personnel. Two such means of upgrading are career counseling and educational opportunities.[31] Although the direct effects of providing a good educational opportunity program and a good career-counseling program are perhaps not as measurable as those of the factors previously discussed, the indirect effects are significant for minorities' upgrading and mobility.

Educational Opportunities

The availability of formal educational opportunities on and off military bases can be a positive factor in the upgrading and mobility of the personnel. This is particularly true for minorities who often have not had the same educational training as nonminorities have had. The armed forces have historically offered service members the opportunity to complete college/vocational training and have made available General Educational

---

[31] The following discussion of both the educational and career counseling systems is based on field interviews, 1974-1975. Field interviews were conducted on board Navy ships and at shore installations in the following cities: San Diego, Calif.; Norfolk, Va.; and Philadelphia, Pa. Interviews were also conducted at the following Marine Corps installations: Quantico, Va.,; and Camp Pendleton, Calif.

Development (GED) programs, which provide members with an opportunity to complete their high school graduation requirements. These programs and others are of special importance to minorities who, because of their educational disadvantage, look to such opportunities as a means of upgrading. Furthermore, educational opportunities such as the above can have a direct impact on an individual's performance and progress in his military job.

Field research of Marine and Navy shore installations revealed well-organized educational services. Normally, complete information was available on all types of educational programs offered. Navy ships, however, displayed some weaknesses which were probably caused by the very nature of shipboard educational programs. Of necessity, the shipboard educational services officer billet is a collateral duty. That is, the officer in charge of educational programs works on a part-time basis. Because of the relatively small number of men on board ships, this billet is not full-time as it is on larger shore installations. As a result, information often is not disseminated, and programs are not formulated in an optimum manner.

The military provides tuition assistance for those successfully completing a course of study in a job-related direction. Much of private industry does likewise. The principal assumption is that the more educated the employee, the more competent and valuable he becomes to the whole work force's effort. Study for educational betterment is, however, placed strictly on an off-duty basis. Naturally, this presents a problem for individuals who wish to continue their education and whose potential class hours conflict with working hours. At some commands, working hours have been arranged around class hours, but such cases are not common. The prospect of adjusting an individual's working hours to accommodate classroom hours is worth considering, as well as the prospect of implementing full-time study and part-time work programs on an individual basis.

Finally, because of congressional cutbacks, the military recently lost GED administration for its members, as well as the United States Armed Forces Institute (USAFI) correspondence course program.[32] The loss of these programs is a tremendous blow to the educational efforts of all services. Perhaps the

---

[32] USAFI provided correspondence courses on a wide range of subjects. In some cases, successful completion of correspondence courses could be used for college, junior college, and high school graduation credit.

Navy has been most profoundly hurt. With a large percentage of total manpower assigned to seagoing units, the phase-out of both GED equivalency tests and of USAFI courses has left a perceptible void. Men attached to seagoing units are usually in port for irregular periods, which militates against their participation in formal, on-shore educational programs.

The BOOST Program

Both the Navy and Marine Corps offer a range of in-service educational programs, many of which lead to a commission in the officer corps. Of particular importance for minorities is the BOOST program (Broadened Opportunity for Officer Selection and Training).[33] BOOST was initially created as a means by which minority representation in the officer corps could be increased. Young minorities who demonstrated potential for programs leading to a commission, such as the Naval Academy or Naval Reserve Officer Training Corps, were selected for a year's course of study in basic subjects, including English, math, physics, and chemistry. After this intensive course of study, it was hoped that the students' scores on standardized tests for admission into these programs would be raised to a competitive level with those applications from the civilian population. The program, conducted in San Diego, has subsequently been opened to all races. In the 1974 class, approximately 65 percent of the students were minorities.

The BOOST program can well be considered an affirmative action program for minorities and other disadvantaged who show potential. It remains one of the only instruments of the Navy and Marine Corps for identifying and helping young sailors and marines with officer potential, particularly minorities, in their bid for a chance for a college education and a commission.

At the program's beginning in 1969, no professional teaching staff was assigned to the school. Rather, the majority of instructors were officers with bachelor degrees and no teaching experience. Therefore, the qualitative results of the program were obviously somewhat dubious in the first years. The program now appears to have a much better teaching staff; many of the officers were previously involved in education or related fields. As the quality of the program has increased, enrollment

---

[33] The BOOST program is fully explained in the Bureau of Naval Personnel, NAVPERS 15791B, Article 1020360.

has also increased. In 1969, enrollment was approximately 35; this figure increased to about 130 in 1973; and expected enrollment in 1978 is approximately 250.

Individuals selected into the BOOST program are not guaranteed an "output"—that is, placement into one of the possible output programs, such as the Naval Academy. As an example, from the class of 1973, only about 40 percent were picked by output programs. Although this program has great potential, BOOST is not being optimally used to provide minorities with a path into the officer corps. It can be argued that, if an individual can perform at a satisfactory level according to appropriate measuring instruments, then after completion of the program, he should be guaranteed a seat in an output program. By this means, both academic and leadership talent can be tapped from within the two services to build a strong, racially balanced officer corps.

Career Counseling

Career counseling, for the purposes of this study, refers to a personnel service designed to provide information on such aspects of the military environment as occupational classification, assignment and promotion opportunities, educational opportunities, advancement procedures, and military law. Although career counseling in the Navy and Marine Corps has been upgraded substantially in recent years, the system of personnel or human resource development could well receive additional emphasis. To be sure, the all-volunteer concept has had a routine effect on increasing the scope, as well as the degree, of commitment made by the military to career counseling, but future efforts to strengthen this service would appear to be a good investment.

Traditionally, the military services have been mission-oriented, with personnel policies being related expressly to mission accomplishment. This practice resulted in intensive training of recruits so that they could spend the greatest balance of their in-service time performing on the job. Dependence on conscription for adequate numbers of personnel, low pay, and many other factors contributed to high personnel turnover rates. The all-volunteer concept has changed the operating premises radically. The military *career* is now something which must be encouraged. The high turnover rates of conscription days are far too wasteful of monetary and manpower resources to be continued.

Good career-counseling capabilities are essential if the services are to compete successfully with the private sector for quality manpower. For each service as a whole, a better career-counseling system should be established within the organization not only for traditional career opportunity information but also for other career-enhancing opportunity information. Such systems now exist, but they are often disorganized, not emphasized, and, therefore, not fully effective. Interviews suggest that there is a general lack of communication and of awareness of career information centers within a command and a general distrust of information received.

As the military organization as a whole becomes more responsive to personnel needs, the individual members of the services, especially those in supervisory and management positions, must become more responsive to the individual needs and goals of those men in their command. This, of course, has direct implications for "striker procedures" and performance evaluation review (see pp. 104-5). It means an increased involvement on the part of senior enlisted and junior officer personnel in counseling and otherwise helping the men under their command to resolve their problems and to realize their goals *within* the military organization.

The effect of an improved career-counseling commitment would be significant for all personnel, especially minority personnel. Under a thorough career-counseling system, service members with real potential or with basic problems can be reached and helped to achieve their goals or resolve their problems within the context of the military organization. The implications for increased personnel performance and higher retention rates are significant.

OFFICER PROMOTIONS

Table VI-3 shows minority officers in both the Navy and the Marine Corps to be concentrated heavily in the lowest ranks. In the Navy, 27.2 percent of black officers are ensigns, compared with 14.3 percent of white officers. In the Marine Corps, 27.3 percent of black officers are second lieutenants, compared with 18.1 percent of white officers.

The immediate question is whether the concentration of minority officers in the lower ranks represents a recent attempt to increase minority officer representation by increasing minority officer accessions or whether it is the result of slower minority officer promotion. Although the question cannot be

TABLE VI-3
*Percentage Distribution of Officer Personnel by
Race and Paygrade in the Navy
and Marine Corps
as of December 31, 1977*

	Navy		Marine Corps	
Paygrade	Total	Black	Total	Black
0-7+	0.4	0.2	0.4	0.0
0-6	5.8	1.8	3.2	0.3
0-5	12.0	3.2	8.0	0.9
0-4	19.7	7.4	14.4	4.6
0-3	27.5	30.5	24.3	19.1
0-2	15.5	21.9	25.5	38.7
0-1	14.3	27.1	18.1	27.3
W1-W4	4.8	7.9	6.1	9.1
TOTAL	100.0	100.0	100.0	100.0

Sources: Bureau of Naval Personnel (Pers-61), "Navy Wide Demographic Data Base for First Quarter FY-78 (1 Oct 77 to 31 Dec 77)" (Washington, D.C.: Department of the Navy, 1978); U.S. Marine Corps, Headquarters, Manpower Planning, Programming and Budgeting Branch, August 1978.

decided conclusively, there is evidence which supports the former possibility as being more correct.

First, it is clear that both the Navy and the Marine Corps have attempted to increase minority representation in their respective officer corps in recent years. As an example, black officer accessions as a percentage of total officer accessions for the Navy had gone from 2.7 in 1970 to 6.5 in 1975. Minorities constituted 2.0 percent of total Marine Corps officer accessions in 1970 and 5.3 percent of total accessions in 1975.[34]

The age distribution of officers also helps to confirm the recent minority accession policies as an explanation of minority concentration in the lower ranks. Forty-four percent of minority officers in the Navy are in the age group twenty-two to twenty-six years, compared with 29 percent of white officers. For the Marine Corps, 46 percent of minority officers are in this age

---

[34] Edward Scarborough, "Minority Participation in the DOD" (Washington, D.C.: Defense Manpower Commission, 1976), Appendix A, Chart XXVIII.

group, compared with 33 percent of white officers. These figures tend to support the contention that minority officers are concentrated in the lower ranks because they are largely recent accessions and have not yet served long enough to be eligible for promotion to the higher ranks.

Although the present concentration of minority officers in the lower ranks appears to be the result of recent attempts to increase minority officer representation, there are certain factors which must be closely monitored to prevent a situation in which minorities fail to advance as quickly as nonminorities. These factors are career pattern (sea/shore rotation, type of positions held, etc.), written fitness reports, and officer warfare specialty (Navy) or military occupational specialty (Marine Corps).

Promotion Standards and Practices

The secretary of the navy annually determines the number of officers to be selected for each rank. Officers are promoted when the actual number of officers in a particular rank (or ranks) falls below the prescribed level. The number of promotions authorized, then, depends upon the expected number of vacancies created by this gap between prescribed and actual numbers.[35] During peacetime, promotion rates slow down, and competition becomes intense for the available spots.

Officers, like enlisted men, have time-in-grade requirements. An officer is normally not eligible to be advanced until he has served in his current rank for a specified period of time. There are exceptions to this rule, however. A maximum of 15 percent of those selected are promoted before reaching their normal eligibility zones ("picked up early"). Those who are eligible but are not promoted face a bleak situation. Only a small percentage of the total number of officers advanced (sometimes only 5 percent) are selected after having been passed over once. It is clear that an officer who has been passed over faces serious problems in trying to make the next higher rank. If, twice successively, an eligible officer is not selected for advancement, then the length of his career is automatically limited.

Selection boards are convened each fiscal year. The boards are composed of senior officers with considerable experience and

---

[35] Bureau of Naval Personnel, Officer Professional Development Division, *Unrestricted Line Officer Career Planning Guidebook*, NAVPERS 15197 (Washington, D.C.: Department of the Navy, 1976), p. 3.

varied backgrounds. The names and records of the officers eligible for selection are submitted to the boards, which review the records and select the officers "best fitted" for promotion. Table VI-4 provides information on the current time-in-grade requirements, as well as the approximate percentage of officers at each paygrade who are eligible for promotion and who are, in fact, promoted.

Although it is not always clear which criteria and factor weights are used in making selections,[36] several factors are carefully considered by each board in determining which officers are best fitted for promotion. Three factors that are consistently important are career pattern, fitness reports, and occupational grouping.

The Importance of Career Pattern

An officer's career pattern will vary with his warfare specialty (or military occupational specialty) both in the Navy and Marine Corps. The career patterns for the Navy's major warfare specialties (surface warfare, nuclear power, and aviation) are different. For illustrative purposes, the surface warfare career pattern is outlined in Figure VI-2. A common feature of this and of all officer career patterns in the Navy and Marine Corps is that early billets are prerequisites for later billets. For example, the first operational tour of a Navy surface warfare officer as a division officer while an ensign and lieutenant junior grade provides the experience, skill, and knowledge necessary to perform the more responsible duties as a department head after making lieutenant on a second operational tour. In short, an officer risks being passed over for the next higher rank if he deviates too far from the "normal" career pattern.

Both the Navy and the Marine Corps have, in recent years, increased their efforts to recruit minorities and both services have instituted race relations/equal opportunity programs. To aid in these efforts, the Navy and Marine Corps have assigned a proportionately large number of minority officers to billets in these areas. This demand for minority officers has fallen most heavily upon minority officers in ranks below lieutenant commander. Although efforts have been made to make recruiting and equal opportunity billets "career enhancing," failure to at-

---

[36] No record is kept of selection board proceedings.

Advancement

TABLE VI-4
Time-in-Grade Requirements for Promotion and Approximate Percentage of Officers in Selection Zone

Rank	Required Years of Commissioned Service Before Eligibility to Given Rank	Approximate Percentage of Officers in Selection Zone Promoted to Rank
Lieutenant Junior Grade	2 years	99
Lieutenant	4 years	95
Lieutenant Commander	9-10 years	80
Commander	15-16 years	70
Captain	21-22 years	60

Source: Bureau of Naval Personnel, Officer Professional Development Division, *Unrestricted Line Officer Career Planning Guidebook*, NAVPERS 15197 (Washington, D.C.: Department of the Navy, 1976).

tain requisite skills and experience by deviating from the normal career pattern early in an officer's career can severely limit his chances of being selected for promotion to commander. And, as mentioned earlier, being passed over once makes it most difficult to be selected later.

The Role of Fitness Reports

Fitness reports are vitally important to an officer's career. Given a normal career pattern, it is often the fitness report that determines whether an officer will be selected for promotion or passed over. Two examples are given in order to demonstrate how subtleties in fitness report writing can make intentional bias felt in an otherwise outstanding fitness report.[37]

A Marine Corps officer had been passed over twice in succession, which in this case meant forced retirement from the service. Upon notification of the second pass-over, he called Marine Corps Headquarters to seek a more detailed reason for the second pass-over. His fitness reports to the best of his knowledge had always been excellent. At the officer's request, his fitness report file was pulled and past fitness reports examined. It was found that the fitness report written prior to the first pass-over decision contained a descriptive paragraph

---

[37] Examples are based on actual situations as reported by Navy and Marine Corps Recruiting Command personnel.

164 Black and Other Minority Participation

FIGURE VI-2
Typical Career Pattern for Surface Warfare Officer

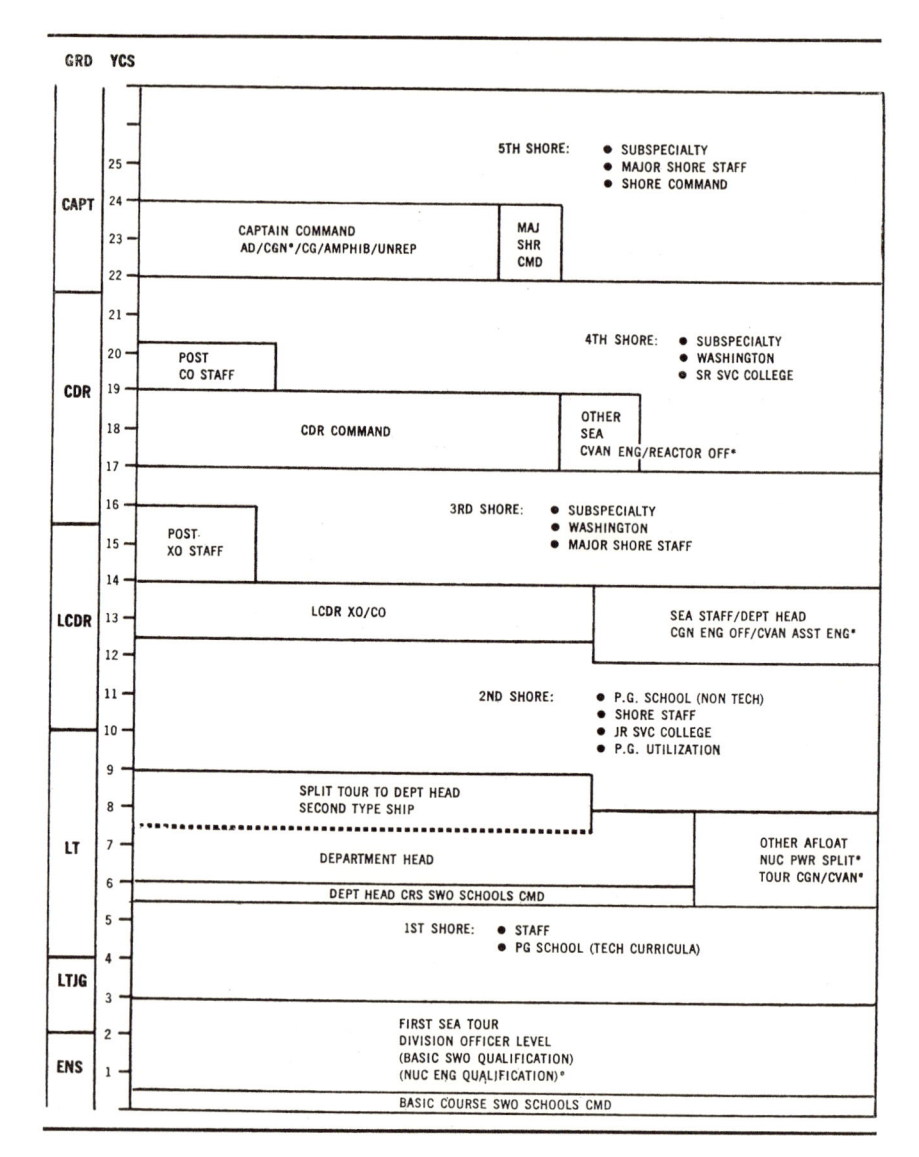

GRD	YCS				
CAPT	25 24 23 22		5TH SHORE:	• SUBSPECIALTY • MAJOR SHORE STAFF • SHORE COMMAND	
		CAPTAIN COMMAND AD/CGN*/CG/AMPHIB/UNREP		MAJ SHR CMD	
CDR	21 20 19 18 17	POST CO STAFF		4TH SHORE:	• SUBSPECIALTY • WASHINGTON • SR SVC COLLEGE
		CDR COMMAND		OTHER SEA CVAN ENG/REACTOR OFF*	
LCDR	16 15 14 13 12	POST XO STAFF	3RD SHORE:	• SUBSPECIALTY • WASHINGTON • MAJOR SHORE STAFF	
		LCDR XO/CO		SEA STAFF/DEPT HEAD CGN ENG OFF/CVAN ASST ENG*	
LT	11 10 9 8 7 6		2ND SHORE:	• P.G. SCHOOL (NON TECH) • SHORE STAFF • JR SVC COLLEGE • P.G. UTILIZATION	
		SPLIT TOUR TO DEPT HEAD SECOND TYPE SHIP			
		DEPARTMENT HEAD		OTHER AFLOAT NUC PWR SPLIT* TOUR CGN/CVAN*	
		DEPT HEAD CRS SWO SCHOOLS CMD			
LTJG	5 4 3		1ST SHORE:	• STAFF • PG SCHOOL (TECH CURRICULA)	
ENS	2 1		FIRST SEA TOUR DIVISION OFFICER LEVEL (BASIC SWO QUALIFICATION) (NUC ENG QUALIFICATION)*		
		BASIC COURSE SWO SCHOOLS CMD			

Source: Bureau of Naval Personnel, Officer Professional Development Division, *Unrestricted Line Officer Career Planning Guidebook*, NAVPERS 15197 (Washington, D.C.: Department of the Navy, 1976), p. 35.

* Nuclear-trained officers.

that was damaging to the officer. This had, in fact, been the reason for the two pass-overs. An investigation proved that the evaluating officer in this case was biased against the individual for personal reasons. The end result was a withdrawal of this fitness report from the officer's file and a removal of all records that he had been passed over for promotion twice. He was subsequently promoted to the next rank.

A Navy officer serving with a destroyer squadron staff received a fitness report that, although rating him in the top 10 percent of officers in his rank Navy-wide, contained a qualification that was damaging to his record. In one portion of the report, the evaluating officer is asked to recommend future jobs for the officer. This officer was not recommended for future staff assignments, which was damning since he was serving on a staff at the time.

The first example points out the inherent danger in the Marine Corps' policy of not showing fitness reports to personnel being evaluated. Both examples serve to show the role of subtleties in fitness report writing. Commanding officers are responsible for officer fitness reports and, certainly in the overwhelming number of cases, try to be as fair and objective as possible in evaluating officer performances. The possibilities of both intentional and unintentional racial bias, along with other biases, do, however, exist.

Warfare Specialty and Military Occupational Specialty

Officer promotion rates tend to be faster in some warfare specialties and military occupational specialties than in others. In the Marine Corps, military occupational specialties that give the officer an opportunity to command men in combat situations generally have higher promotion rates than other specialties. In the Navy, nuclear-qualified officers, particularly those serving on board nuclear-powered submarines, tend to be promoted more quickly and regularly than other officers.

The barriers to minorities' participation in some specialties were discussed in chapter V. It is important to note that minority officers are disproportionately concentrated in occupational classifications which do not offer command opportunities; those officers are therefore at a competitive disadvantage. Furthermore, minorities, as compared with nonminorities, face barriers to selection for specialties, such as nuclear power, in which promotion tends to be faster and more regular.

CHAPTER VII

Retention

The military services are seeking to build and maintain professional career forces. Retention of competent personnel is a key element in reaching this goal. The Navy and the Marine Corps, as well as the other services, invest considerable time, effort, and money to induce their personnel to remain in service for a career.

Retention is also an important issue for minorities' participation in the services. The degree to which minorities are retained has an important impact, of course, on their representation in the military. Furthermore, minority personnel retention is a factor in determining the number of minorities available and eligible for promotion to the higher paygrades and ranks within the career forces. The purposes of this chapter are, first, to bring into focus the policies and practices of the Navy and Marine Corps with respect to the retention of officer and enlisted personnel and, second, to analyze these policies and procedures with regard to the retention of minorities.

DEFINITION OF CAREER

In the Navy and Marine Corps, an enlisted man is considered *career* when he has reenlisted at least once. The normal initial term of enlistment for enlisted personnel is four years. There are, however, circumstances under which an individual can initially enlist for an obligated period of service of up to six years.[1]

Similarly, a career officer is an officer serving beyond his initial period of obligated service. Officers enter the service for an ini-

---

[1] The Navy has a five-year enlistment contract. Both the Navy and Marine Corps have a six-year contract. These programs are small relative to the four-year program. Individuals who enlist under the six-year contract are assigned to extremely technical training in occupations such as advanced electronics or nuclear power. Navy Recruiting Command, *Navy Recruiting Manual—Enlisted*, COMNAVCRUITCOM Instruction 1130.8A (Washington, D.C.: Department of the Navy, 1975), p. 26-1.

tial period, usually three to five years, during which time they are obligated to remain in service. Subsequent to this initial obligation, the officer "serves at the pleasure of the president," which generally means that an officer may serve as long as he likes, provided that he maintains an acceptable performance record and receives normal promotions. Normally, he may leave whenever he chooses, pending approval of a resignation request which he must submit.

Although the definitions for *career* and *retention* are those now used by the services, they are not universally accepted. The Defense Manpower Commission in its report to the president and the Congress stated that the term *career* should only be applied to individuals with a high probability of remaining in the service until retirement. The commission recommended that the term *career* should only be applied "after completion of ten years of total active Federal military service. . . ."[2]

REENLISTMENT ELIGIBILITY

Meeting the eligibility requirements is the first step for individual service members in the reenlistment process. The enlisted marine must have no reenlistment restrictions in his contract and must not show a negative trend in his disciplinary record or have committed any serious offenses. For a first reenlistment, he must not have been convicted of a court-martial offense or have been awarded nonjudicial punishment more than twice. The individual marine must meet Marine Corps appearance and physical standards and must have achieved certain minimum conduct and proficiency marks. The individual sailor must meet similar standards of discipline and performance, as well as the Navy's physical requirements. Finally, for an individual to qualify for retention in either the Navy or Marine Corps, he must be recommended by his commanding officer.[3]

In addition, each service has "up-or-out" promotion criteria that career personnel must meet in order to reenlist. For example, Marine corporals and below may not reenlist for a period

---

[2] Defense Manpower Commission, *Defense Manpower: The Keystone of National Security* (Washington, D.C.: U.S. Government Printing Office, 1976), p. 257.

[3] Bureau of Naval Personnel, "Reenlistment Quality Control Program," BUPERS Instruction 1133.22E (Washington, D.C.: Department of the Navy, 1976).

resulting in more than ten years of total active service. If an individual has not been promoted higher than the rank of corporal after eight years, then he is not eligible for reenlistment. Sergeants, except those who have not yet been considered for promotion, may not reenlist for a period resulting in more than twelve years' total service. Sergeants who have failed promotion twice are not eligible to reenlist without the Marine Corps commandant's approval.[4]

A first reenlistment in the Navy requires a sailor to advance to E-4 or above, or at least to have passed the E-4 examination without having been advanced. A sailor who makes the grade of E-3 during a first enlistment may request a one-time two-year extension, during which time he must make E-4 in order to reenlist. Sailors who have reenlisted one or more times are entitled to reenlist as long as their time in service does not exceed a maximum number of years which is determined by their paygrades. The following schedule shows this relationship:[5]

Paygrade Attained	May reenlist, not to exceed a total service time (years) of
E-4	20
E-5	21
E-6	21 or 24 (depends on Career Reenlistment Objectives [CREO] rating
E-7	27
E-8, E-9	30

There are those who take issue with the up-or-out policies practiced by the services because of their emphasis on screening members out of the career force rather than managing the flow into the career force. It is argued that the up-or-out policies punish the failure to advance, but do not reward achievement. Up or out conveys the message to service members who are not selected for advancement that they can no longer contribute to the mission. Nevertheless, "it is inconceivable that a Service member who has been screened many times during his Service

---

[4] U.S. Marine Corps, Headquarters, *Career Planning and Development Guide*, Vol. I, *Administration*, Marine Corps Order P1040.31A (Washington, D.C., 1977), Section 3004.

[5] Bureau of Naval Personnel, BUPERS Instruction 1133.22E.

Retention

life by other promotion boards, by Service school and other selection boards, and by other evaluations is suddenly of no further value to his Service simply because the Service does not have enough promotions to go around."[6]

Yet a retention process could perhaps be established by which the services consider more closely each member's worth to the organization. It becomes more and more important for any individual to develop leadership skills as he is promoted and, therefore, finds himself in more of a supervisory role in his occupational specialty. Personnel who demonstrate a good technical grasp of their job but who are either poor or marginal administrators are penalized and often not recommended for advancement by their commanding officers. It would seem that there is room for such individuals in a position requiring less administrative skill. A retention process which bases its decision upon both a member's contribution and his promotion rate could accommodate these individuals.

Reenlistment Eligibility vs. Occupational Specialty Manning Levels

The Navy's system of up-or-out reenlistment regulations contains an additional complex formula which was implemented in January 1976. In brief, the formula combines a sailor's length of service, occupational specialty, and subspecialty with the needs of the service for career personnel in various occupations and paygrades. The formula firmly ties reenlistment rules to the CREO program.[7] Not only, then, is reenlistment eligibility determined by an individual's promotion rate, but now is also determined by how well or poorly his occupation is manned with career personnel. Therefore, reenlistment eligibility criteria have shifted more towards the needs of the Navy and promotion opportunities within an occupational specialty.

If an individual requests to reenlist in his current occupational specialty and has satisfactory promotion progress, per-

---

[6] Defense Manpower Commission, *Defense Manpower: The Keystone of National Security*, p. 261.

[7] Career Reenlistment Objectives is a list of open, neutral, and closed promotion potential occupations. Open occupations are those with less than 89 percent career manning. Neutral occupations are those with 90-105 percent career manning. Closed occupational specialties are those with more than 105 percent career manning. Bureau of Naval Personnel, "Career Reenlistment Objectives (CREO)," BUPERS Instruction 1133.25C (Washington, D.C.: Department of the Navy, 1975).

formance evaluations, etc., then the manning level of his occupational specialty becomes the primary consideration. Those who are in a specialty that is either properly manned or undermanned at the higher enlisted paygrades are granted reenlistment in their specialty as a matter of course, but those who desire to reenlist in an occupation that is overmanned may face difficulties. The term for which these individuals can reenlist may be limited, depending on manning levels, their demonstrated ability to progress, and their total years of service. But it is by no means clear that the Navy will allow persons in overmanned categories to remain in their current occupations. Other alternatives include requiring an individual to convert to another less crowded occupation. Or an individual might be required, if he has the minimum service time necessary, to transfer to fleet reserve. Finally, an individual might be refused permission to reenlist. This is particularly possible for personnel who have demonstrated only a marginal promotion rate.[8]

The new CREO modifications to the reenlistment system are a distinct improvement over the old method of selection. The new rules should work to the individual's advantage, as well as to the Navy's. An individual can be shifted to another occupational specialty in which there is a better chance for advancement. There are programs, which will be discussed shortly, that provide the mechanism for this shift to another occupational specialty. Extra schooling is possible for personnel being shifted so that transfers can be made laterally. At present, however, an individual aspiring to an undermanned occupation requiring formal school training must, generally, meet the Basic Test Battery (BTB) score requirements for that occupation. For those who satisfy this requirement, not only is the advancement opportunity likely to be better in the less crowded fields, but there is also the possibility that they will qualify for other incentives, such as reenlistment bonuses or proficiency pay.

Unfortunately, minorities are disproportionately represented in the overmanned or poorest promotion rate occupations. Furthermore, minorities face a severe barrier to reenlistment-induced occupational transfer. Transfer to a more critical specialty requires being able to score high enough on the BTB to qualify for the formal training. And, as previously discussed, minorities,

---

[8] Jim Parker, "POs Get Tough 'Up-or-Out' Rules," *Navy Times*, December 3, 1975.

as a group, score significantly lower on these tests than nonminorities score. As the reenlistment transfer requirements now stand, little can be expected for large-scale transferring of minorities to better opportunity specialties.

MINORITY VS. NONMINORITY RETENTION SITUATION: STATISTICAL ANALYSIS

Time in the Navy has a very strong positive relationship with paygrade level attainment. The samples used in chapter IV from which the original preservice plus in-service, cross-sectional, and performance models were derived represent different time frames. Because the time frames are different, the sample paygrade distributions are different. Because of the frequency distribution of time in service, the original model describes the advancement function most reliably from 1973 to 1975. On the other hand, everyone in the cross-sectional model served in the Navy as an enlisted man between 1971 and 1975 as of the file date. This sample consists of individuals who have four years of active service. Interpretations are based on the situation which existed during this specific four-year period. In the performance evaluation model, only paygrades E-5 and E-6 are well represented, whereas paygrades E-1 through E-6 are well represented in the original and cross-sectional models. Furthermore, the performance evaluation model describes the enlisted advancement function during a different time frame than is described in either of the other two models. The original and cross-sectional samples have a median value for time in the Navy of twenty-one months and four years, respectively. In the performance evaluation model, the mean and median values for time in the Navy are both seventy-one months. So it can be seen that we are looking not only at a more specific paygrade group than in the other models but also at a different time frame.

Interpretation of Model Results with respect to Minority Personnel Retention

The period between 1971 and 1975 saw considerable activity in the formulation of minority-upgrading policy and manpower-planning policy. The armed forces reduced their manpower requirements considerably in an effort to scale down after Vietnam. The Navy established minority personnel recruiting and upgrading goals, including occupational classification priorities for qualified minorities. The original sample frequency distribution of time in service by race-ethnic group provides insight into the emphasis the Navy has placed on minority personnel recruiting. For example, 10.6 percent and 7.8 percent of the preservice plus in-service sample are black and other minorities, respectively. Approximately 50.0 percent of these minorities had lengths of service of less than or equal to twelve months as of the file date. When this is compared with 28.0 percent of the aggregate sample, or somewhat less than that for nonminorities having equivalent lengths of service, the Navy's recent minority-recruiting effort is brought into focus.

If, however, the past rates of attrition of minorities and nonminorities reflect the situation to be expected in the future, proportionate minority representation will not be possible despite increased recruiting efforts. The median time in service for members of this sample was approximately six years, and all were in paygrades E-5 and above. From the model results, it is concluded that the attrition rate for minorities is higher, and that most minorities who enter the Navy leave sooner and are less likely to advance to E-5 or above than their nonminority counterparts.

What factors explain the disproportionate minority personnel attrition rates during the period evaluated by the models? The scaling down of military manpower requirements took place between 1972 and 1975. As a result, individuals were discharged prior to completion of obligated service. The results of the statistical analysis clearly indicate a higher attrition rate for those individuals having lower academic credentials than for those having higher academic credentials. Given the educationally disadvantaged status of minorities, it is not surprising that they had a disproportionately high attrition rate.

The differences in minorities' and nonminorities' promotion rates can also be examined as a function of length of service. Because promotion rates are tied to reenlistment up-or-out policy, it is of interest to determine if group promotion differences remain constant as time in service increases. When the significant preservice and in-service variables are controlled for, minorities are found to be promoted slightly more slowly than nonminorities during their first two years of active duty. This conclusion is based on the results of the original preservice plus in-service model. It was found in the aggregate that, given a minority member and a majority member who entered the Navy in 1973, completed two years of active service, and have identical academic credentials, marriage status, time in paygrade, discipline record, and occupational classification characteristics, the minority member will be promoted at a slightly slower rate than the majority member. On the other hand, the results of the cross-sectional and performance evaluation models indicate that those minorities who are eligible to reenlist and do remain in the military after their initial four-year obligation are promoted at least as quickly as those nonminorities with identical significant variable characteristics.

Minority Personnel Representation vs. Reenlistment Eligibility

Table VII-1 shows Navy retention statistics for the third quarter of 1976. The figures are in percent for first-term and career enlisted personnel by race. The "eligible" figures represent the percentage of enlisted who meet the reenlistment eligibility criteria outlined earlier. The "reenlisted" figures show the percentage of those eligible who actually reenlist. The survivability rate shows the percentage of all enlisted who meet the eligibility requirements and reenlist.

For first-term personnel, there are significant differences across the racial groupings in the percentage eligible to reenlist. The percentage of nonminorities who are eligible to reenlist is 11.1 percent higher than the percentage of blacks who are eligible to reenlist. The percentage of those eligible to reenlist that actually do reenlist also shows some differences across racial groupings. It is of interest to note that blacks who are eligible to reenlist do so at a rate 4.0 percent higher than nonminorities. The net result, as can be seen in Table VII-1, is that 18.6 percent of nonminorities and 17.1 percent of blacks meet the eligibility requirements and reenlist.

Career personnel show similar trends. Again, the survivability rate of blacks is not significantly different from that of nonminorities. Concerning both the first-term and career retention statistics, a smaller percentage of blacks are eligible to reenlist than nonminorities, but relative to nonminorities, a larger percentage of blacks who are eligible to reenlist actually do so. This is especially true among career personnel. It is clear that the retention problem with respect to blacks involves their inability to meet reenlistment eligibility requirements. Given the high reenlistment rate of eligible blacks, compared with that of

TABLE VII-1
Enlisted Retention and Reenlistment Race-Ethnic Comparisons
July 1 through September 30, 1976

Category	Caucasian	Black
First Term		
Eligible	60.0%	48.9%
Reenlisted [a]	31.0	35.0
Survivability Rate [b]	18.6	17.1
Career		
Eligible	64.5	56.8
Reenlisted [a]	61.9	78.8
Survivability Rate [b]	39.9	44.7

Source: Bureau of Naval Personnel, MAPMIS 1133-2251 (Washington, D.C.: Department of the Navy, September 30, 1976).
Note: The data are based on DOD formula for unadjusted reenlistment rate.
[a] Percentage of those eligible.
[b] Percent eligible x percent reenlisted.

eligible whites, the Navy could significantly increase minority representation by concentrating on helping blacks meet the eligibility requirements.

REENLISTMENT INCENTIVES

Once reenlistment eligibility has been determined, the individual must decide whether to reenlist. The services invest much money, time, and effort in order to encourage personnel reenlistment. Each service has both monetary and nonmonetary incentives. The nonmonetary incentives consist of training, specialty change, and duty station guarantees.

The single monetary incentive for both services is a Department of Defense authorized program called Selective Reenlistment Bonus (SRB). The SRB authorizes enlisted personnel to receive an additional fractional multiple of their basic pay upon reenlistment or extension of present enlistment for at least three years. The multiple by which pay is increased varies with job category, changes in the job over the years, and the career manning percentages in the job category. In other words, the objective of the bonus is to "increase the number of reenlistments in those ratings characterized by retention levels insufficient to man adequately the career force."[9]

The Marine Corps' nonmonetary reenlistment incentives are quite similar to the Navy's. These incentives can include one or more of the following: formal school training for those qualified to enter a formal school; a change in military occupational specialty for occupations not requiring formal school training; a choice of duty station or type of duty; a guarantee of not being transferred from present duty station; and, finally, promotion not to exceed the rank of sergeant, granted upon recommendation by the individual's commanding officer.[10] The above incentives are outlined in the Marine Corps' *Career Planning and Development Guide*, but not in the detail provided by the relevant Bureau of Naval Personnel instruction.[11] Because the Navy's and Marine Corps' nonmonetary reenlistment incentives of formal training and lateral transfer are similar, only the Navy programs are described and analyzed here.

Selective Conversion and Reenlistment (SCORE) Program

The SCORE program offers qualified individuals an opportunity to transfer from one occupational specialty to another as a reenlistment inducement. The Navy benefits by such a program by being able to shift manpower in accordance with need. The individual also benefits, as stated before, through being placed in an occupation with greater promotion potential.

As set forth in the *Bureau of Naval Personnel Manual*, the SCORE program contains the following incentives:[12]

1. guaranteed assignment to formal school (Class A) training[13] with automatic transfer to the appropriate oc-

---

[9] Bureau of Naval Personnel, "Selective Reenlistment Bonus Program," BUPERS Instruction 1133.26A (Washington, D.C.: Department of the Navy, 1976), p. 1.

[10] U.S. Marine Corps, Headquarters, Marine Corps Order P1040.31A.

[11] Bureau of Naval Personnel, BUPERS Instruction 1133.22E.

[12] Bureau of Naval Personnel, *Bureau of Naval Personnel Manual*, NAVPERS 15791B (Washington, D.C.: Department of the Navy, 1969), Article 1060010.

[13] Class A school is the *initial* formal training required to enter certain ratings (occupational specialties).

cupational specialty upon satisfactory completion of the school curriculum;
2. guaranteed assignment to higher level formal school (Class C) training;[14]
3. automatic advancement to petty officer second class (E-5) upon satisfactory completion of the Class C school curriculum;
4. reenlistment bonus if eligible;
5. eligibility to reenlist more than one year early after successful completion of Class A school.

For a member to be eligible to take part in the SCORE program, he must

1. be currently assigned to an overmanned or neutral specialty;
2. be a petty officer first class (E-6), petty officer second class (E-5), petty officer third class (E-4), or an E-3 who is also a designated striker;
3. meet minimum formal school Basic Test Battery (BTB) score requirements;
4. demonstrate potential for the desired occupation, have superior performance evaluations, and be highly recommended by the commanding officer;
5. not have been awarded nonjudicial punishment more than once for a period of 18 months preceding the date of application and not have been convicted of a court-martial or civil offense on the current enlistment or within 48 months preceding the date of application;
6. have at least 21 months continuous active Navy service, but not more than 15 years of total military service;
7. be within four months of completion of minimum active duty tour;
8. not have received benefits from the Selective Training and Reenlistment program;

---

[14] Class C school is advanced training for personnel already serving in a particular rating. A service member choosing to reenlist under the SCORE program may receive guaranteed assignments to both A and C schools providing he is qualified for both. Since some C schools have minimum time-in-rating requirements, a service member may be required to return to operational duty between A school and C school.

9. agree to extend for transfer of occupational specialties and to enlist or reenlist for a total obligated service of six years after transfer.[15]

The requirements that an individual must meet to be eligible for a transfer can be stringent. Of particular significance for the Navy's ability to achieve its equal opportunity objectives is the effect of the BTB score requirements. Although it is possible to obtain a waiver of this requirement with a carefully documented recommendation from the individual's commanding officer, the exam cut-off scores generally serve to preclude a significant redistribution of minority personnel to the more technical occupations.

Selective Training and Reenlistment (STAR) Program

Another program designed to encourage reenlistments is the STAR program. This program serves as a reenlistment incentive through its guarantee of formal school training for those who qualify. The SCORE program provides an individual with an opportunity to attend formal school training while being laterally transferred to a new occupational specialty. The STAR program does not provide for such a transfer, but does provide for formal schooling in the individual's current specialty.

The following reenlistment incentives are provided by the STAR program:

1. guaranteed assignment to Class A or Class C formal school;
2. guaranteed advancement to petty officer second class (E-5) upon completion of Class C school;
3. reenlistment bonus if eligible.[16]

The eligibility requirements for the STAR program are that an applicant must

1. be currently in a CREO group other than that which represents occupations which are most overmanned, or be in a critical skill subspecialty;

---

[15] Bureau of Naval Personnel, NAVPERS 15791B, Article 1060010.

[16] *Ibid.*, Article 1060020.

2. be recommended by his commanding officer for career designation, demonstrate above average career potential, and meet considerably higher than the minimum reenlistment standards;

3. be a petty officer second class (E-5), petty officer third class (E-4), or an E-3 and meet the required promotion rate standards;

4. have at least 21 months, but not more than five years, continuous Navy service and not more than eight years total active military service;

5. reenlist for six years;

6. meet minimum test score requirements for formal school (Class A) training;

7. not have received a court-martial conviction nor have been awarded nonjudicial punishment more than once for 18 months preceding the date of application;

8. meet minimum obligated service requirement upon entry into the guaranteed school;

9. not have derived any benefits from the SCORE program.[17]

Both the STAR and the SCORE programs provide for formal school training, but both programs are somewhat limited in their ability to provide widespread reenlistment incentives because individuals taking part must meet the BTB score requirements. For the most part, those who meet these test requirements have already attended formal schools during their first enlistment and are located in the better promotion potential specialties. Although commanding officers can make recommendations for test score waivers in individual cases, the heavy weighting of exam scores makes it difficult to use either the STAR or SCORE programs as a means to upgrade the educationally disadvantaged. Therefore, the current programs' ability to upgrade minorities is limited.

Both STAR and SCORE have the *potential* of expanding the opportunities available for those who demonstrate ability through on-the-job performance, but lack educational experience. If the Navy is to utilize this potential fully, however, it is imperative

---

[17] *Ibid.*

that on-the-job performance be weighted more and BTB scores be weighted less in the selection process, or that means be provided by which the educational experience can be attained. In doing so, the result will be an increased manpower pool which is eligible for these programs. Therefore, some modification of the selection criteria will provide advantages for the Navy, as well as for those who are capable, but educationally disadvantaged. The Navy is an organization which suffers from overmanning in some specialties and critical manpower shortages in others. If the selection criteria for the SCORE program are modified, then the Navy will have more flexibility in its ability to redistribute career personnel from overmanned to undermanned occupational specialties.

With the implementation of the proposed modification, it is recommended that the SCORE and STAR programs be expanded. It is not suggested here that the personnel qualifications be lowered, but rather that the selection criteria for these programs be expanded so that a "whole-man" evaluation be made as a matter of course. If an individual has demonstrated his ability to the satisfaction of his commanding officer, then that fact should have weight in the selection process. An individual's on-the-job performance should not be relegated to a mere prerequisite for further consideration. Rather, it should be given a specified weight and combined with other relevant variables to make a good selection decision. In cases in which an individual is deemed to have the potential for a job, but lacks the education to complete formal training successfully, remedial education could be provided to the advantage of the Navy, as well as of the individual.

Guaranteed Assignment Retention Detailing (GUARD) Program

Enlisted personnel with nuclear training and experience are in great demand in the Navy. For this reason, they are not normally eligible for a rating change and have already received considerable formal training. The STAR and SCORE programs do not, therefore, apply to nuclear-trained personnel. There is, however, a reenlistment incentive program designed specifically for nuclear-trained personnel—Guaranteed Assignment Retention Detailing (GUARD).

According to the Navy's *Enlisted Transfer Manual*, GUARD is for nuclear-trained personnel with six to ten years' service in ratings electronics technician, electrican's mate, interior communications specialist, machinist's mate, engineman, and boiler technician. The program provides that those serving in nuclear billets at sea may request a guaranteed shore assignment in the port of their choice with an option for a guaranteed follow-on assignment. Additionally, a waiver of up to twelve months of the projected sea tour may be granted. The program also provides those with nuclear training currently serving ashore with the opportunity to request a nuclear-powered ship in the homeport of their choice.[18] As noted, few minorities have qualified for nuclear experience.

OFFICER RETENTION

As noted at the beginning of this chapter, an officer incurs an initial service requirement upon commissioning and thereafter serves at the pleasure of the president (with one exception, which is described on page 181). Because officers do not normally incur a second period of obligated service, many officers serve for short periods beyond their initial service requirement. Approximately 80 percent of all officers who leave the service without completing a twenty-year career do so within two years of their initial service requirement.[19] Retention of an officer, therefore, is considered to have taken place if he is still in the service two years after his initial service requirement has been completed.

In order to remain eligible for continued service, an officer must not be passed over two consecutive times for promotion to the next higher rank.[20] Promotion is, therefore, essentially the single criterion for an officer's maintaining service eligibility. For the purposes of this section, it is sufficient to recall that three factors heavily affect an officer's promotion: career pattern (sea/shore rotation and type of positions held), written performance evaluations, and officer designator (military occupational specialty).

[18] Bureau of Naval Personnel, *Enlisted Transfer Manual*, NAVPERS 15909B (Washington, D.C.: Department of the Navy, 1967), Section 8.02.

[19] Bureau of Naval Personnel (Pers-402b), "Officer Personnel Retention Statistics," CNO/SECNAV Point Paper (Washington, D.C.: Department of the Navy, July 15, 1976).

[20] If an officer reaches the rank of lieutenant commander (0-4), he can normally expect to serve for a full twenty years before the possibility of a second pass-over forces his retirement.

Officer Incentives

The Navy pays what can be described as a reenlistment cash bonus to nuclear-trained officers as an incentive to remain in the Navy. Nuclear-trained officers also receive additional monthly payments, as do aviators, medical doctors, and other nonnuclear-designated submariners.[21] As noted, few minorities have qualified for these posts.

The retention of nuclear-trained officers is an important personnel problem in the modern Navy. The Navy invests tremendous sums to train officers for qualification in nuclear power. After training and qualification, however, officers in the nuclear power program, whether surface or submarine warfare designated, have skills that are highly marketable in the private sector. The Navy, therefore, offers nuclear-qualified officers who are beyond their initial service requirement a bonus of up to $20,000 if they agree to obligate themselves for four more years of service. In addition, all officers, nuclear-trained, as well as others, serving on submarines receive additional monthly pay called submarine pay. Submarine pay was paid in the past to compensate personnel for the additional hazards and discomforts of submarine duty. Today, with larger and relatively safe submarines, this additional pay is recognized as incentive pay.

The Navy also spends large sums training pilots and naval flight officers (NFO). This training gives the pilot or NFO highly marketable skills. In addition, aviation, particularly carrier-based aviation, is somewhat more hazardous than other warfare specialties. In order to compensate and retain pilots and NFOs, the Navy gives aviation personnel additional monthly pay called aviation officer incentive pay. The Marine Corps also has an aviation officer incentive pay program which is similar to that of the Navy. There are no other Marine Corps officer incentive pay programs.

MINORITY OFFICER RETENTION

Table VII-2 shows the retention rate of Navy officers in each of the four major warfare communities for fiscal years 1970 through 1976. The percentage figures represent the fraction of

---

[21] Department of Defense, *Military Pay and Allowance Manual* (Washington, D.C.: Department of Defense, n.d.), Part Three, chapter 1.

TABLE VII-2
Navy Officer Retention by Warfare Community

Warfare Community	FY71	FY72	FY73	FY74	FY75	FY76	Steady State Goals
Nuclear Submarine	33	41	47	a	41	36	51
Pilots	27	34	43	44	45	52	52
Naval Flight Officers	51	42	40	49	47	54	43
Surface [b]	17	14	14	14	30	39	30

Sources: Bureau of Naval Personnel (Pers-402d), "Officer Personnel Retention Statistics," CNO/SECNAV Point Paper (Washington, D.C.: Department of the Navy, June 15, 1974); Bureau of Naval Personnel (Pers-402b), "Officer Personnel Retention Statistics," CNO/SECNAV Point Paper (Washington, D.C.: Department of the Navy, July 15, 1976).

Note: Numbers represent the percentage of officers remaining on active duty who are two or more years beyond their initial service requirement.

[a] FY74 statistics do not include USNA graduates commissioned in FY68, because of an increase in service obligations which began with that graduating class. The effect of no USNA graduates is greater in the nuclear submarine community where USNA graduates would otherwise have composed approximately one-half of the base of the FY74 statistics.

[b] Commencing in FY75, surface includes only those officers designated 111x. Previously, it included those officers designated 110x or 111x.

officers who, for each fiscal year, are two or more years beyond their initial service requirements and are still on active duty. The steady state retention goals set by the Navy are also given.

Minority officer retention statistics can be confusing and misleading. The difference of only a few minority officer resignations during any period can have a relatively large effect on the retention percentage. Despite the lack of large amounts of meaningful data, some inferences can be made about present and future trends in minority officer retention.

As shown in Table VII-2, officer retention is considerably higher in the nuclear submarine and aviation communities than in the surface community. There are almost no minority officers serving in the nuclear power Navy. Additionally, there are very few minorities in the Navy and Marine Corps aviation programs. Only 18.4 percent of all black officers and 11.1 percent of all other minority officers are in aviation in the Navy, compared with 30.8 percent of all nonminority officers. As discussed in

Retention

chapter V, minority officer attrition from the aviation training programs is disproportionately high, and minority entrance into the nuclear power training program is still almost nonexistent.

Minority officer attrition generally is also high. Two factors appear to account for this. First, minority officers are not assigned to the occupational specialties that have the highest retention rates. The facts that retention in the Navy is highest in the aviation and nuclear submarine communities and that minority officers tend to be concentrated elsewhere help to explain the high minority officer attrition rates. As the Navy increases its efforts to recruit minority officers, an effort must also be made to facilitate their entry into the aviation and nuclear communities. Failure to do so will continue to have an impact on the retention of minority officers.

Second, the demand for college-educated minorities in private industry is very high. Minority officers in the armed forces are, of course, predominantly college-educated. Therefore, they are very much in demand. In order to retain them, the military must compete with private industry, which, faced with affirmative action requirements, is anxious to place minorities in management positions. A minority person with experience as a Navy or Marine Corps officer usually has abundant job opportunities outside the military.

Minority Officer Career Eligibility

Because promotion rate is the predominant criterion in maintaining retention eligibility, the promotion variables which have been obstacles for minority officer advancement also have an adverse impact on minority officer retention. Another point of interest is the variance in officer loss rates among the various procurement sources. With the exception of physicians and dentists procured by direct appointment, the "highest loss rate is among officers who were commissioned through OCS (officer candidate school) programs."[22] It was pointed out in chapter III that the Navy and Marine Corps rely heavily on OCS programs as a source of minority officers. Over 50 percent of the Navy's black officers and more than 40 percent of the Marine Corps' black officers are procured through OCS programs. It is clear that more emphasis must be placed on routing qualified

---

[22] Edward Scarborough, "Minority Participation in the DOD" (Washington, D.C.: Defense Manpower Commission, 1976), p. 30.

minorities into the officer candidate programs which have a better retention record, particularly the U.S. Naval Academy.

Finally, the impact of performance evaluations on minority officers' eligibility rates, as compared with those of nonminority officers, has now been evaluated. H. Minton Francis, deputy assistant secretary of defense, has made public the results of a 1975 computer analysis showing that black officers consistently received performance marks five to ten points lower than their white counterparts.[23] These evaluations, used as a basis for periodic reductions in the services' officer corps and a key factor in promotion, affect retention. The present trend of lower performance marks for minority officers has clear and obvious implications for future minority officer retention. It is extremely important that racial bias not enter into these subjective evaluations. The results of the above study indicate that the need for continued effort and, possibly, *increased* emphasis on the services' human relations programs are needed.

---

[23] Lionel Bascom, "Unwelcome in the Military," *Philadelphia Inquirer*, July 17, 1976, p. 4-A.

CHAPTER VIII

Concluding Remarks

Executive Order 11246 requires employers operating under its jurisdiction to avoid discrimination and to take affirmative action to ensure the absence of discrimination in all phases of employment. The requirements for an affirmative action program include the obligation to conduct an analysis of minority personnel utilization deficiencies and to pursue actively the implementation of minority utilization goals and corresponding timetables. In 1970, Executive Order 11246 and its implementing regulations were extended to cover the Navy and Marine Corps, as well as the other armed services.

The services have established goals and objectives to provide both for equal opportunity and for affirmative action programs. In addition to seeking to eliminate racial bias, the Navy and Marine Corps have committed themselves to distributing minorities proportionately across paygrade and rank categories, as well as across occupational groups. Furthermore, the services have committed themselves to increasing the total number of minorities in service, especially in the officer corps.

Today, the Marine corps has a more than proportionate percentage of minorities, the Navy a deficiency in that ratio. Both services, like private industry, have had difficulty in finding minorities for their officer and higher skilled positions.

What are the prospects for the Navy's and Marine Corps' achieving more nearly proportionate representation in the technical occupations, higher enlisted paygrades, and all specialties and ranks of the officer corps? Based on the results of this study, it is clear that the prospects in the near future are not good. First of all, the supply of such personnel in the labor market is not sufficient to achieve such equality, especially in view of the armed services' forced competition in the labor market with private industry for that supply. In addition, although the services have made great strides in the last ten years in increasing the percentage of minorities in the enlisted ranks, not enough has been done to upgrade these minority members nor to recruit and retain minority officers. The results of the study further indicate that, if the operating institutional policies and procedures now in effect remain in effect, there will be little, if any, improvement in the minority-upgrading situation for many years to come.

RECRUITING ORGANIZATION

Before the advent of the all-volunteer armed forces, the recruiting of personnel for the Navy and the Marine Corps was a relatively simple operation. The Navy and Marine Corps were able to fill their ranks to a great extent with young men eagerly seeking an alternative to Army service. By simply placing induction points or recruiting stations at locations near heavy concentrations of draft-age young men, the services could count on a stable stream of manpower.

Similarly, prior to the present requirements to meet minority-recruiting goals, little consideration had to be given to the special problems of minority recruiting. There was almost no special attention given to the recruiting of minorities with service-eligible qualifications; certainly no consideration was given to the proportionate representation philosophy under which the services are now working. Today, both the all-volunteer armed forces and affirmative action requirements are realities. Despite these changes in the recruiting environment, the services' recruitment policies and procedures, to a large extent, have not been modified to accommodate either the existence of the all-volunteer force or the idea of affirmative action.

It is clear that the Navy and Marine Corps must place additional emphasis on recruiting techniques and effective salesmanship in general. An effective recruiter force must be maintained in order to compete with private industry for the most qualified personnel. Concerning minorities, increased emphasis must be placed on minority-recruiting problems and community penetration techniques. Professional recruiters need to be trained and utilized, current manpower research explored, and rewards developed in the system for effective recruiting. Minority recruiting could be enhanced by utilizing minority recruiters and more thorough techniques for penetrating the minority community. Much could be learned from private industry in this regard.

Concluding Remarks

RECRUITMENT AND CLASSIFICATION ELIGIBILITY: MENTAL APTITUDE TESTING

Mental aptitude tests are used to predict both an individual's success in the service and his success in a particular occupation. For recruitment consideration, an individual's mental aptitude test score and educational level jointly determine eligibility. An individual must score above a certain cut-off score to be further considered for recruitment. Occupational classification is based, for practical purposes, strictly on mental aptitude test performance. Like the recruitment eligibility exam requirement, cut-off scores are also established for formal training schools and, therefore, for various occupational specialties. An individual is eligible for a specific formal school only if he scores above the cut-off score required for entrance into that school. For both recruitment and classification, physical, moral, and other personal characteristics are used only as disqualifiers. Operationally, then, the mental aptitude exams serve either as the sole personnel selection criterion or the most heavily weighted criterion in determining occupational classification and recruitment eligibility, respectively. Because minorities, as a group, do not score as well as nonminorities on these exams, they have been, and continue to be, disproportionately denied enlistment and relegated to the nonskilled, closed occupations.

The aptitude tests appear to have an inflationary impact on standards and should be reexamined in view of the propensity for cut-off scores to vary with manpower needs. There still is insufficient information whether the present apparently inflated standards are the most cost-effective. If they are not, a key basis for relying on inflated test criteria which eliminate a sizable proportion of minorities is not proper.

Because the mental aptitude exam operationally serves as the predominant, if not sole, criterion for recruitment and classification, an individual's moral and physical attributes, as well as other personal characteristics and background, are considered only as potential disqualifiers. A recruitment or classification decision based solely on academic credentials and moral and physical disqualifiers is, at best, incomplete. Additional measures of an individual's performance capability must be developed. Incorporation of reliable measures of factors such as cooperativeness, reliability, initiative, conduct, vocational interest, attitude, and resourcefulness into the selection decision would add to the ability to measure performance potential. Experience should also be considered. This additional information would allow the armed services to place less emphasis on academic credentials as measures of performance potential. More emphasis would then be placed on past nonacademic performance. Perhaps something like assessment centers could be developed. The assessment center method has been used to detect management potential in blue-collar workers. The researchers have reported success in measuring personal attributes such as creativity, human relations skills, behavior flexibility, leadership, stress resistance, and other factors. A methodology similar to that of the assessment center, which provides an *expanded* measure of an individual's capabilities and needs, should be considered by the military for enlisted personnel recruitment and classification.

There can be no doubt that the services' current recruitment and formal training selection policies provide dim prospects for significant minority upgrading in the foreseeable future. Yet, it must be said that the Navy and Marine Corps have made some effort to mitigate the situation. Both services now consider an individual's race-ethnic characteristics in determining classification. This has helped to distribute minorities who are deemed qualified to the more technical, fast promotion track occupations. Nevertheless, an individual must still qualify under inflated mental standards for a particular formal training program and, thus, for the corresponding occupations. Because a disproportionate percentage of minorities do not qualify for technical training, this race-ethnic consideration has been only marginally successful.

Academic credentials must be reexamined and further justified if minority utilization is to be enhanced. But this must also be combined with remedial education so that appropriate academic standards can be met.

REMEDIAL EDUCATION

The military provides disadvantaged personnel with a means to upgrade themselves. It is true, traditionally, that the military has served as a positive social institution through its recruitment and enlistment of members of disadvantaged groups within the country. Many of the nation's recognized minority leaders credit much of their success and the development of their leadership and management skills to military experience. It is also certainly fair to say that many of the minorities who are now in the skilled labor force developed those skills while in the military. Our research, however, requires us to question whether the services are living up to their full potential as a vocational training institution for their disadvantaged personnel.

Because of the heavily weighted mental aptitude system used to match individuals with jobs, the educationally disadvantaged entering the military have been, and are being, relegated to nonskill jobs in which there is little or no formal training. Those entering the military with the highest academic credentials are being assigned to formal school training and afterward to a technical occupation. These individuals are certainly afforded an opportunity for upgrading. Yet, those who enter the military and are educationally and/or vocationally disadvantaged are offered, under most circumstances, no opportunity for formal skill training or skilled occupation classification. Rather, they are assigned to soft-skill jobs. Unfortunately, minorities are disproportionately represented in this category. For individuals assigned to soft-skill jobs, there is little educational or vocational benefit to be gained from joining the military. Furthermore, promotion rates for these members are slower than for individuals in the more technical occupations. And, finally, post-service employment prospects are much worse for persons who do not learn a technical skill while in the service.

Upgrading the educationally disadvantaged, if it is to take place on a large scale, will almost certainly be realized through an effective national manpower program. There are good economic, as well as social, reasons why our nation's institutions, both governmental and nongovernmental, should better develop, deploy, and utilize our human resources. It is imperative that a national manpower policy involve all sectors of the economy, including all levels of government, in order to be fully effective. It must be recognized that the military, as the largest vocational training institution in the United States, has the potential to contribute significantly toward manpower policy. The ability of the services to upgrade minorities may well be inextricably tied to their commitment to a progressive manpower policy and, thus, to affirmative action.

It is strongly recommended here that an effective remedial education program be developed by the armed services to provide *capable,* yet educationally disadvantaged, personnel with the opportunity to complete formal school training successfully.

Through a remedial program, these individuals can be upgraded. This is good human resource management. A successful remedial program will increase the productivity level, advancement rate, and postservice employment opportunities of those disadvantaged persons who participate. It is noted that, because the program will provide for a more productive individual, the military will benefit, as well as the individual, through that individual's increased contribution to the organization. Because minorities represent a disproportionate percentage of the educationally disadvantaged group, a remedial education program designed to upgrade the disadvantaged will automatically affect a large number of minorities in the services. The Job Oriented Basic Skills (JOBS) program to be initiated jointly by the Department of Health, Education, and Welfare and the Navy is a fresh start and is viewed with optimism.

The question of whether or not it is cost-effective to upgrade the educationally disadvantaged remains. As stated, the armed services have, historically, sent to formal school training only those who score highest on the mental aptitude exams. This method of formal school selection was designed for the high input/high turnover draft-motivated military. The argument for not providing a remedial education program is that the cost of upgrading individuals is an unnecessary expense additional to formal training. The all-volunteer force, however, might very well require a different classification philosophy. It is true that the Navy and Marine Corps are currently filling their formal school quotas with qualified personnel without incurring the additional costs. But can this already qualified manpower pool be counted on as the economy expands? Furthermore, it is quite possible that higher aptitude individuals are no longer less costly overall, if indeed they ever were. It is hypothesized here that remedial education for those showing potential will prove cost-effective by virtue of increased productivity and retention. Research is needed to determine the validity of this hypothesis.

Whatever remedial plan is devised, if additional funding is required, its implementation will rest with Congress. Recent congressional actions have been hostile to all programs that the military is conducting which are not directly related to national defense. Congressional cutbacks have reached into many military educational programs already. The military services are aware of the growing problem of basic educational deficiencies in today's recruit, and they correctly see this as an impediment to effective military force performance. Congress, as the ultimate supplier of funds, must be cognizant of this situation and its implications for effective manpower utilization and the ability of the military to achieve its mission. The all-volunteer concept places the services in direct competition with private industry in the labor market. Today's recruit is looking for training and the opportunity for advancement, and if he lacks the basic skills necessary, the services must help him to realize his objectives. If not, the traditional high input/high turnover rates of the military services will continue. Because minorities form a significant proportion of the service members who would be affected by remedial programs, the potential effect on their upgrading and mobility is substantial.

From a strategic vantage point, Congress should consider the military's potential for upgrading *capable,* but disadvantaged, personnel as a viable means of helping to achieve national manpower objectives. The Department of Defense has conducted an enormous amount of human resource development research. This research has concerned personnel-upgrading problems such as performance measurement, alternative testing techniques, test validation, vocational training measures, culture-fair testing, and differential validity.

With respect to personnel development research, in many areas the Department of Defense is at the forefront. Furthermore, the Department of Defense has the resources to develop and implement personnel-upgrading programs. Although much research has been conducted, implementation of such programs has been slow because of congressional funding. Because developing job skills and providing remedial education for competent, yet disadvantaged, personnel is not the primary mission of the armed services, funding has been hard to get. Yet, the military can assume a leadership role in such activities without compromising its primary mission. Furthermore, the military may very well be forced to provide these services to individuals in order to compete with the private sector for manpower and to use efficiently those who do enlist. Therefore, perhaps, for the long-term viability of the all-volunteer system and because of the Department of Defense's personnel research capability and its ability to implement personnel-upgrading programs, the military should be seriously considered as a major forum for personnel upgrading. Financing for such programs could then be at least partially shifted away from public works jobs and private industry subsidizing. Because of the controlled military environment, monitoring of the system would be more manageable.

ALTERNATIVE TRAINING METHODS: SELF-PACED

The self-paced training method provides another means through which the participation rate of competent, yet disadvantaged, personnel in the formal training programs can be increased. It provides for increased minority participation through modification of the traditional method of formal school training. Self-paced training allows less emphasis to be placed on mental aptitude testing. Variable course completion time frames are established for individuals enrolled in selected formal schools. The time allotted for a recruit to finish his formal schooling is a function of that individual's aptitude credentials. Self-paced learning allows individuals with different mental classifications to take part in the same formal training. Those with higher examination selector scores are expected to complete training more quickly than those with lower selector scores. Reward, punishment, and remedial systems are set up to ensure the timely course completion by each individual according to his expected time frame for course completion.

The Navy has begun using the self-paced learning method for many of its formal training courses. Research is ongoing, and indications are that the program has great potential. The average student, having the current formal school academic credentials requirements, is successfully completing the necessary course work in one-third less time. Preliminary findings indicate that this method, first, is justified by its cost-effectiveness and, second, provides strong potential for the upgrading of educationally disadvantaged personnel.

Because of the differential time frame in self-paced learning, *less* emphasis on mental aptitude is required. The services must reevaluate their selector score system and perhaps replace it with one which takes into account a potential student's motivation and familiarity with the subject matter. The implications of this system for minorities and nonminorities who could not qualify for such training are significant. Full utilization and expansion of the self-paced training program are fully endorsed.

UNDESIGNATED STRIKERS AND ON-THE-JOB TRAINING

Recruits who are assigned to a command directly from boot camp are categorized as undesignated strikers. Undesignated strikers have no formal school occupational training and are normally placed by the operational command into nonskilled occupations in accordance with the needs of the command. Often, individuals who report directly to commands from boot camp have not qualified for formal school training by virtue of their relatively poor performance on the aptitude examinations.

In both the Navy and Marine Corps, approximately 40 percent of recruit accessions are assigned to commands as undesignated strikers. Minorities are disproportionately represented in occupations in which persons report directly to operational commands as undesignated strikers. Therefore, the policies and procedures with respect to undesignated striker job selection and training are important considerations of this study.

We found that the placement of undesignated strikers is haphazard at best and frequently results in a disgruntled employee or one who loses a chance for a meaningful career. Accordingly, we recommend that the placement of these individuals be more carefully handled so that they are provided information on opportunities and cannot be utilized solely to fill miscellaneous and unskilled manpower slots.

PROMOTION ELIGIBILITY

The Navy and Marine Corps agree that the selection criteria for promotion should be based on the "whole-man" concept. Each service, however, emphasizes different factors in its promotion decision. The most significant difference between selection criteria used by these services is that the Navy uses advancement examinations as a heavily weighted factor in the promotion decision, while the Marine Corps does not use written examinations to such a great extent. The Navy advancement exams are used to provide a measure of one's theoretical and applied knowledge of his occupation.

Prior to 1974, Navy advancement exams were found, in many cases, to be the only variable considered in the promotion decision because of excessively high cut-off scores. This was so despite official policy which called for consideration of other factors as well. Since 1974, the promotion decision mechanism has moved toward an expanded factor composite in selecting individuals for promotion and has thus downgraded somewhat the factor weighting of the advancement exam. Ostensibly, the Navy now weights these exams in accordance with policy. The Navy claims that the system selects for promotion those candidates who have the best overall mix of technical knowledge, on-the-job performance, and experience.

Advancement Exam Weighting

As in the recruitment and classification system, the advancement examination is the most heavily weighted factor in the promotion decision. Furthermore, not only is the advancement examination the predominant factor, but again, as with recruitment and classification policy, an individual must score above a cut-off score on the exam in order to remain in contention for promotion.

The rationale put forth for requiring cut-off scores is that the Navy does not want to promote individuals who do not have an adequate technical background in their occupational specialty. Yet, the argument can be made, as it was in our discussion of recruitment and classification, that there has been no historically demonstrated concern for ensuring that the technical knowledge required for advancement is above a stipulated minimum score. Instead, the cut-off scores have been determined somewhat arbitrarily. They have been set in order to fill quotas and, therefore, have fluctuated as a function of the manpower pool. There has been no research conducted to indicate at what point the cut-off score should be set. It must again be emphatically stated that, if the Navy insists on using cut-off scores, they should be validated. That is, research must be conducted to provide reliable evidence that those not scoring above the designated cut-off score are incapable of performing at the next higher paygrade satisfactorily.

Recent research also challenges the hypothesis that the advancement exam is the most reliable predictor of who will perform best at the next higher paygrade. If the results of this research are found to be the general rule, then the Navy's rationale for heavily weighting the exam is contradicted, and a strong case can be made for placing additional emphasis on on-the-job evaluation variables.

Furthermore, in many instances, the best performers are not the best test-takers. Although all personnel allowed to take the exam have been recommended for promotion by their commanding officers, it is quite possible for those personnel who have not performed as well as others in the past to be advanced, while those who have demonstrated stronger on-the-job performance are not advanced.

Alternative Policy for Promotion Selection

The Navy promotion system needs to be improved. It is imperative that research focus on developing a system which selects personnel for promotion according to the best measure of on-the-job performance at the next higher paygrade. Within this context, the whole-man concept of evaluation must be fully developed. It is also necessary that the system be evolved within the framework of the Navy's equal opportunity objectives.

Admittedly, it is no easy task to construct a "good" method of predicting on-the-job performance at the next higher paygrade and to provide minorities with increased opportunity for advancement. There is evidence, however, that suggests that both objectives can be partially realized by downgrading the weight of the advancement exam, modifying the examination cut-off score requirement, and increasing the weight of on-the-job performance evaluations.

Of the various types of selection actions, selection for promotion can offer one of the best potentials for improved opportunity in career growth for minorities. The services' promotion systems have undergone some fundamental modifications during the last few years. In general, there has been a shifting toward a truer whole-man concept of evaluation. It has been argued here that the Navy, especially, has not done enough to develop its promotion function along these lines. Both the Navy and Marine Corps are increasing their emphasis on on-the-job performance as a means to demonstrate potential for greater responsibility. Within this conceptual framework, the importance of an individual's performance evaluation has increased and should continue to do so.

It is not proposed here that technical knowledge evaluations be abandoned. On the contrary, there is most definitely a need to test an individual's technical knowledge of his field when he is being considered for promotion. Two important questions remain unanswered, however, with respect to the role of advancement exams in the promotion process. First, what weight should be assigned to these examinations? Second, how should an individual's technical expertise be ascertained? Further research must be conducted in order to answer these questions. Perhaps the alternative testing research being conducted will provide a means for achieving culture-fair testing.

ALTERNATIVE TESTING

The armed services have conducted, and continue to conduct, a great deal of research on personnel selection. Much effort has focused on the development of new culture-fair tests that might be used in place of the current examinations. Similar-item-difficulty, computerized, and job-sample tests are three methods of personnel selection that are being studied by the Navy. Of the three methods, the job-sample test offers the most viable means of alternative testing. It is a fresh approach taken to confront the problem of culture-fair testing.

The job-sample test is a practical exam composed of representative samples of the work involved in the job for which an individual is being considered. It is not a traditional paper-and-pencil test. Rather, it is a nonverbal, seemingly fair testing method. Of all tests in the employment-selection and promotion field, these job-sample measures, because they require the skills demanded in a given job, are generally considered by critics of testing to be least objectionable. The Navy is conducting research on job-sample testing to determine whether demonstrated ability to learn selected aspects of a job can be employed as a predictor of ability to learn to perform the total job.

How well do job-sample tests predict performance on the job? This question has not been answered conclusively. Research is ongoing. The Navy's initial results do indicate that job-sample testing can be useful as a personnel selection device, especially for manually oriented occupations. The validity of this method of measurement for occupations requiring abstract reasoning has yet to be determined. It is quite possible, though, that job-sample tests will be found more reliable in their prediction of on-the-job performance than the traditional paper-and-pencil exams for some occupations. This remains to be seen. Initial evidence does indicate that, when this form of measurement is used in the selection process, the advancement potential of minority personnel is increased.

PERFORMANCE EVALUATIONS

Performance evaluations play an extremely important role in the promotion process. Performance on the job, to a large extent, determines whether or not an individual is recommended for promotion. Recall that a recommendation for promotion is a prerequisite for further promotion consideration. Furthermore, these evaluations determine an individual's reenlistment eligibility.

It is through these subjective evaluations that direct bias can affect an individual's ability to compete with his peers for advancement. The procedure is certainly not without weaknesses. Indeed, a key factor in any system of evaluation is that the individual being evaluated knows how the system operates, its significance for his advancement, and, most important, the contents of the periodic evaluation report. In this manner, the individual is made aware of his performance for the preceding period and, it is hoped, will receive deserved praise or constructive criticism. In the Navy, this process is accomplished by requiring the evaluated individual's signature on the report itself, plus a face-to-face consultation between the individual and his immediate superior to review and discuss the evaluation marks.

The Marine Corps diverges markedly from the Navy in this practice. The Marine Corps maintains an official policy of not showing marines (officer *or* enlisted) their fitness reports unless they are derogatory, in which case the individual must sign the report. It is a fact that the wording of descriptive paragraphs within the evaluation report is so crucial that the insertion of a "but" or "however" or other qualifiers may remove an individual from further consideration for advancement. Such phraseology, however, may not be considered "derogatory" by the evaluator. The Marine Corps' policy of not requiring hard-nosed counseling of an individual on his performance evaluation unless that member has performed badly is obviously in need of revision. It is clear that the potential for the abuse of all service members, not just minority personnel, exists, for the system permits bias, including racial discrimination, and eliminates potentially valuable counseling sessions. The benefits of such periodic one-to-one counseling sessions are obvious. Not only does the individual know exactly where he stands, but he also has the opportunity to receive personal counseling which will, it is hoped, yield a better marine.

The services' promotion systems have undergone some fundamental modifications in recent years. The Navy and Marine Corps have increased their emphasis on on-the-job performance as a measure of one's ability to perform successfully at the next higher paygrade. From the results of this study, it is recommended that the military continue to shift more weight to measures of on-the-job performance. Within this framework, the importance of an individual's performance evaluation has increased. As the on-the-job evaluations become more and more important, it will become even more crucial that all members understand the function and importance of these evaluations, the periodic performance reviews, and the counseling sessions. A good human relations environment will also become increasingly important to ensure that racial bias does not influence the promotion decision through on-the-job performance evaluations.

MILITARY JUSTICE

The statistical analysis results discussed in chapter IV indicate that an individual's disciplinary record strongly influences his promotion opportunities. Unfortunately, a disproportionate share of minorities, relative to nonminorities, were found to be reduced in paygrade during the course of their enlistment. Because disciplinary action determines, to a large extent, whether or not an individual is advanced, the impact of more than proportionate "busts" on minorities' advancement is severe.

The causes of the disparate offense rates between race-ethnic groups are not clear. It has not been conclusively determined whether or not these differences are due to institutional or interpersonal racial discrimination. The nonjudicial punishment (NJP) system is a target of such charges. Recent research, however, has found no differences in disciplinary actions across race-ethnic groups for similar offenses.

Interesting results that provide some insight into the problem have been derived from research investigating the relationship between race-ethnic groups' NJP rates and a command's organizational effectiveness. NJP rates are strongly related to how well the personnel thinks its command manages human resources (organizational effectiveness). With respect to such factors as "communications flow" and "supervisory support," blacks perceive their command's environment less favorably than nonminorities perceive it. Moreover, the similarity or difference between the

Concluding Remarks 199

race of an individual and the race of his supervisor has a significant effect on an individual's perception of his command's human resource management capability.

The implication is that a serious problem appears to exist between enlisted blacks and their immediate supervisors, particularly when the supervisor is white. It appears that blacks feel that they are not being supported by their supervisors, and that they are being left out of formal communications channels; it appears that the friction at this supervisor/subordinate level may be leading to excessive black NJP rates. It is recalled that an evaluation of the Navy's Phase I Race Relations Program found that senior enlisted supervisors were least affected by the race relations training. These findings strongly suggest that additional emphasis should be placed on the services' race relations programs. Furthermore, perhaps special effort should be directed at the worker/supervisor relationship.

CAREER COUNSELING

Career counseling, for the purposes of this study, is defined as a personnel service designed to provide information on such aspects of the military environment as occupational classification, assignment, promotion opportunities, educational opportunities, advancement procedures, and military law.

It is fair to say that, traditionally, the military services have been mission-oriented, with the only attention to personnel policies being related expressly to mission accomplishment. As already stated, this practice resulted in intensive training of recruits so that they could spend the greatest balance of their in-service time performing on the job. Dependence on conscription for adequate numbers of personnel, low pay, and many other factors contributed to very high personnel turnover rates. The all-volunteer concept has changed the operating premises radically. The military *career* is now something that must be encouraged. The high turnover rates of conscription days are far too wasteful of monetary and manpower resources to be continued.

Good career-counseling capabilities are a must if the services are to compete successfully with the private sector for quality manpower. For each service as a whole, a better career-counseling system must be established within the organization not only for traditional career opportunity information but also for other career-enhancing opportunity information. Such systems now exist, but they are often disorganized, not emphasized, and, therefore, not fully effective. Interviews that were conducted suggest that there is a general lack of communication and of awareness of career information centers within a command and a general distrust of information received.

The effect of an improved career-counseling commitment would be beneficial for the personnel, especially minority personnel. Under a thorough career-counseling system, service members with real potential or with basic problems could be reached and helped to achieve their goals or resolve their problems within the context of the military organization. The implications for increased personnel performance and higher retention rates are significant.

RETENTION

The military services are seeking to build and maintain professional career forces. The retention of competent personnel is a key element in reaching this goal. The Navy and the Marine Corps, as well as the other services, invest considerable time, effort, and money to induce their personnel to remain in service for a career.

Retention is also an important issue for minority utilization. The degree to which minorities are retained has an important impact, of course, on their representation in the military. Furthermore, minority retention is a factor in determining the number of minorities available and eligible for promotion to the higher paygrades and ranks within the career forces.

Reenlistment Eligibility

Although minorities, if eligible, are more likely to reenlist than eligible nonminorities, a much larger percentage of minorities, relative to nonminorities, do not qualify for reenlistment. On balance, in the Navy the overall minority reenlistment rate is lower than the overall reenlistment rate for nonminority personnel.

The differences in minorities' and nonminorities' reenlistment rates were briefly discussed in terms of discipline record, performance evaluations, paygrade level attainment, and occupational classification. The significance of all four factors in determining reenlistment eligibility affects minorities' reenlistment opportunities negatively. That is, minorities, as a group, are more often disqualified for reenlistment by any one of the above factors than are nonminorities. Of the factors that can exclude an individual

Concluding Remarks

from reenlistment eligibility, paygrade level attainment and occupational classification are the two that account for most of the cases in which servicemen are deemed ineligible.

Each service has "up-or-out" promotion criteria that career personnel must meet in order to reenlist. Minorities are disproportionately deemed ineligible for reenlistment as a result of the promotion criterion. The up-or-out requirements have, however, been somewhat deemphasized recently. Therefore, the promotion criterion no longer constitutes the reenlistment eligibility barrier that it was in the past.

Lateral Transfer and Reenlistment-Induced Formal Training

The new modifications of the reenlistment system are a distinct improvement over the old method of selection. The new rules should work to the individual's advantage, as well as to the Navy's. An individual can be shifted to another occupational specialty in which there is a better chance for advancement. There are programs that provide the mechanism for this shift to another occupational specialty. Extra schooling is possible for personnel being shifted so that a transfer can be made laterally. At present, however, an individual aspiring to an undermanned occupation requiring formal school training must, generally, meet the mental aptitude score requirements for that occupation. For those who satisfy this requirement, not only is the advancement opportunity likely to be better in the less crowded fields, but there is also the possibility that they will qualify for other incentives, such as reenlistment bonuses or proficiency pay.

Unfortunately, minorities are disproportionately represented in the overmanned, or poorest promotion rate, occupations. Furthermore, minorities face a severe barrier to reenlistment-induced occupational transfer; transfer to a more critical specialty requires being able to score high enough on the mental aptitude exams to qualify for the formal training. For the most part, those who meet these test requirements have already attended formal schools during their first enlistment and are located in the better promotion potential specialties. Although commanding officers can make recommendations for test score waivers in individual cases, the heavy weighting of exam scores makes it difficult to use reenlistment-induced formal training programs as a means to upgrade the educationally disadvantaged. Therefore, as the reenlistment transfer requirements now stand, little can be expected for large-scale transferring of minorities to better opportunity specialties.

The lateral transfer and reenlistment-induced transfer programs do have the *potential* of expanding the opportunities available for those who demonstrate ability on the job, but lack educational experience. If the Navy is fully to realize this potential, however, it is imperative that on-the-job performance be weighted more and aptitude scores be weighted less in the selection process, or that remedial training be provided. In doing so, the result will be an increased manpower pool which is eligible for these programs. Therefore, the above recommended modification of selection criteria will provide advantages for the Navy, as well as for those in service who are capable, but are educationally disadvantaged. The Navy is an organization that suffers from overmanning in some specialties and critical manpower shortages in others. If the selection criteria for the transfer programs are modified, then the Navy will have more flexibility in its ability to redistribute career personnel from overmanned to undermanned occupational specialties.

With the implementation of the proposed modification, it is recommended that the lateral transfer and reenlistment-induced formal training programs be expanded. It is not suggested here that the personnel qualifications be lowered, but rather that the selection criteria for these programs be expanded so that a wholeman evaluation be made as a matter of course. If an individual has demonstrated his ability to the satisfaction of his commanding officer, then that fact should have weight in the program selection decision-making process. An individual's on-the-job performance should not be relegated to a mere prerequisite for further consideration. Rather, it should be given a specified weight and combined with other relevant variables to make a good selection decision.

OFFICERS

Historically, minority entrance into the officer corps of the armed services has been even more restricted than their entrance into the enlisted ranks. For much of the first half of the twentieth century, minorities in the Navy were restricted to service in the enlisted ranks as stewards, and very few found their way into the officer corps. As late as 1946, no blacks served as Marine Corps officers. Moreover, minorities have never approached proportionate representation across the officer ranks and occupational specialties in either the Navy or the Marine Corps.

Concluding Remarks

Minority Officer Recruiting

As of June 1977, blacks constituted only 1.8 percent and 3.6 percent of the officer corps of the Navy and Marine Corps, respectively. Clearly, an enormous recruiting effort is required by each service if minorities are ever to approach proportionate representation.

The minority-recruiting task is complicated by a number of factors, including the general requirement that officers have college degrees. Although some officers have "come up through the ranks," most enter the services at the officer level after successfully completing a college education. Minority entrance into college degree programs has increased over the years, yet there remains a dearth of minorities holding such degrees. As a result, the services are forced to compete for officer candidates from a very small manpower pool.

The Navy and the Marine Corps rely heavily on their respective Officer Candidate School (OCS) programs to train minority officers. OCS programs in turn rely heavily on the recruiting organization to induce college graduates to enter the service. Hence, the Navy and the Marine Corps are competing in the labor market for already graduated minorities. Competitors in this small manpower market include private sector firms capable of offering attractive compensation packages and more progressive affirmative action incentives. Public sector agencies at all levels are also aggressively competing for college-educated minority personnel. Such competition obviously restricts the services in their attempts to procure an increasing number of minority officers.

The logical alternative for the Navy and Marine Corps is to seek commitments from promising minority personnel *prior* to college graduation by offering them Naval Reserve Officer Training Corps scholarships and Naval Academy appointments. By attacking the "pregraduate" market, the Navy and Marine Corps could avoid the aggressive competition for already graduated minorities. Techniques exist and are being experimented with by the Navy for identification of potential candidates at the high school and junior college level. Effective minority community penetration programs may be a logical means to identify, groom, and interest potential candidates in becoming military officers. The Navy has had success with a college preparatory program for enlisted personnel by which a large percentage of those graduating from the program are channeled into officer-training programs such as NROTC or the United States Naval Academy.

Officer Classification and Assignment

Minority officers are disproportionately underrepresented across officer occupational specialties. Barriers exist that tend to prevent minority participation in some of the more prestigious occupations, such as aviation and nuclear power, in which incentive pay and/or greater promotional opportunities exist. These barriers include stringent entrance standards and high minority attrition from training programs.

The services' aviation programs require candidates to pass a battery of written examinations before being admitted into flight training. Minority officers face the same problems with these examinations that enlisted minorities face with the mental aptitude examinations.

Minority officers who enter the services' flight program have an unusually high attrition rate. The program is rigorous, requiring engineering and mathematical skills. The fact that a much smaller percentage of minorities than of nonminorities entering the flight program have the needed mathematical and "hard science" skills is believed to contribute to the disproportionate minority attrition rate. Furthermore, according to the flight instructors interviewed, minorities experience more difficulty with the predominant teaching method—programmed texts.

Minority officer participation in the Navy's nuclear power program has been almost nonexistent. The chief factor in this regard has been the academic standards used in selection. Nearly all candidates selected for nuclear power training have had undergraduate or graduate education in the physical sciences and/or engineering. Moreover, in recent years, nearly one-half of the selections have come from among Naval Academy graduates. Relatively few minorities possess these requisite credentials.

The best course of action for the Navy in trying to increase minority entrance into the nuclear power and aviation fields is to identify and interest promising minority candidates in becoming Naval officers and provide these individuals with a rigorous engineering-related college education, such as that provided by the United States Naval Academy.

Officer Promotions

Minority officers in both the Navy and the Marine Corps are heavily concentrated in the lowest ranks. Available evidence,

Concluding Remarks

however, indicates that this concentration is attributable to recent attempts by the services to increase minority representation in their officer corps. The greater proportion of minority officers have not been in service long enough to have been considered for promotion into the higher ranks. There are, nevertheless, three important factors that must be closely monitored to ensure that future promotion patterns are equitable. These are career pattern, fitness reports, and occupational specialty.

Promotion depends upon the ability of an officer to fill the more responsible operational billets at the higher levels within his occupational specialty. Early billets are prerequisites for the later, more responsible billets. Hence, failure to fill the prerequisite billets early in an officer's career—that is, deviating from career pattern—is detrimental to his promotion opportunities. The services must ensure that minority officers are given ample opportunity to stay within their chosen career patterns. In the recent past, extensive use has been made of minority junior officers in filling equal opportunity and recruiting billets. Although considered to be necessary in meeting affirmative action and equal opportunity objectives, this practice, in many cases, has forced minority officers to deviate from their career patterns. The services must make every effort to balance the need for minority officers in recruiting and equal opportunity billets with career pattern demands, which are so important for promotion consideration.

Fitness reports are perhaps the single most important factor in promotion consideration for officers. Yet, available data indicate that commanding officers have tended to give minority officers less favorable fitness reports than nonminority officers. Whether this is because of generally poorer minority performance, bias in fitness report writing, or both cannot be conclusively determined from available data. Nevertheless, it is quite clear that the subjective and subtle nature of fitness report writing is susceptible to bias, racial or otherwise. The Marine Corps' policy of not showing officers their fitness reports compounds further the issue of fairness.

An officer's occupational specialty also has an impact on promotion opportunity. Promotion rates tend to be faster in some specialties than in others. It is clear that the services must make an effort to ensure that minority officers are adequately represented in those specialties in which promotion opportunity is greatest.

Officer Retention

Because of the small number of minority officers in the Navy and the Marine Corps, it is difficult to determine conclusively whether or not minority officers face higher attrition rates than nonminority officers. There are, however, indicators which suggest that the minority officer attrition rate is higher.

Promotion is the key to officer retention. The services are reminded here that the same factors which tend to limit minority officers' advancement also have a negative impact on their retention.

Another key retention factor is the source through which the officer is procured. Attrition is highest among officers who enter the services through OCS programs. The Navy and the Marine Corps, as noted, currently rely heavily on OCS programs for minority officer recruitment. Hence, as well as having a negative impact on the services' ability to recruit minority officers, the heavy reliance on OCS programs also tends to reduce minority officer retention.

In the Navy, officer retention tends to be higher in the aviation and nuclear submarine communities than in the surface warfare community. It should also be noted that aviators and submariners receive incentive pay; surface officers do not. Minority officers, as noted, are more heavily concentrated in the surface warfare community than in either the aviation or submarine communities. Again, this minority officer occupational distribution has a negative impact on minority officer retention.

FINAL COMMENT

In addition to preserving our land in time of emergency, the armed services have made other equally great contributions to our society. In attempting to meet the pressing problem of social and economic equality, the services have done much, but can make even a greater contribution than they have heretofore. Our recommendations for improvements in the policies of the Navy and Marine Corps are designed to strengthen that contribution without in any way weakening those services' basic mission or their resolve to accomplish that mission.

Appendix A

*The Multiple Stepwise Regression Models:
Statistical Technique*

A stepwise analysis was used to identify which variables from the group of explanatory variables available should be used in the regression model. The stepwise analysis technique provided a means of screening the explanatory variables based upon an evaluation of the variables with respect to their relationship with the dependent variable. This technique provided insight into the relative strengths of these relationships between the proposed independent variables and the dependent variable.

A program from the Biomedical Computer Programs (BMD)[1] package BMDP2R, was used to estimate the parameters of the multiple regression equations in a stepwise manner. The so-called F method stepping algorithm was selected from four possible stepping algorithms. The F method allowed the computer program to move from one regression iteration to the next while evaluating the regression equations. Each iteration provided the opportunity for another variable to enter the regression equation according to the following rule: If one or more available variables were out of the regression equation, the one having the highest F value entered the equation if it passed the tolerance test. That is, the explanatory variable outside the equation with the highest computed F value entered the equation if that value was greater than the F-to-enter limit used in the program. The standard F-to-enter limit suggested by the program, namely 4.0, was used.

As already stated, the multiple stepwise regression model provided information on the relative contribution of the explanatory variables in explaining the paygrade (dependent variable) reached by an individual. The program used started with an independent variable and added another one during each iteration. It stopped making iterations when no remaining variables had a computer F value greater than or equal to the F-to-enter limit.

An F-to-enter value of 4.0 required the relationship between the dependent variable and the combination of explanatory variables to be significant. The F-to-enter value of 4.0 required, furthermore, that each of the independent variables entering the equation be significantly related with paygrade at least at the 0.05 level of significance. The F-to-enter value of 4.0 was, however, low enough to allow the independent variables which increased the coefficient of multiple determination (R^2) only slight-

---

[1] W. J. Dixon, ed., *Biomedical Computer Programs* (Berkeley, Calif.: University of California Press, 1975), p. 305.

Appendix A

ly to enter the regression equation. This value provided for a less stringent means of screening variables than a higher F-to-enter value. A higher F-to-enter value criterion would have allowed only the *most* significant variables to enter the regression equation. But, by using the standard value, the researcher was able to evaluate the relative importance and contribution of a wider range of statistically significant variables.

The BMD program provided the statistics to determine whether or not the regression equation was "good." This information was generated during each iteration. The multiple correlation existing between the dependent and independent variables (R^2) was calculated in order to determine how good the regression equation was. The program also computed the regression coefficients of the variables entering the sequential regression equation, as well as their F-to-enter values. Finally, a simple correlation matrix was provided by the program.

The multiple correlation (R^2) measures the proportion of the total variation of the dependent variable which is explained by the regression equation. A high R^2 was sought. The higher the R^2, the greater the success of the regression equation in explaining the variation of paygrade level attainment. Because these models were used to obtain insight into minority-upgrading problems, rather than to predict paygrade, the value of R^2 was viewed in conjunction with the other statistics provided by the program. The simple correlation coefficients and the regression coefficients added to the interpretation of the results as well.

The statistical significance of each explanatory variable was determined by an F ratio. The computed value of F is the ratio between the additional variance explained by the addition of each independent variable and the unexplained variance. Tests of significance were performed for each independent variable by comparing the computed F value with the critical F value in the same manner as for the test of overall relationship discussed above. That is, a significant relationship existed between an independent variable and the dependent variable if the computed F value was greater than the critical F value. The F-to-enter criterion of 4.0 ensured that the variables included in the regression model were significant at the 0.05 level of significance.

The program computed the regression coefficients for the significant explanatory variables. In a geometric sense, a regression coefficient represents the slope of the resulting straight line in the plane described by the dependent variable and the corresponding independent variable while holding the other independent variables constant. It is an estimate, obtained from the studied sample, of the unknown population coefficient. The final regression coefficient for a particular independent variable is affected by the other significant independent variables. This is so because the coefficient measures the contribution of the variable in defining the slope of the final regression line which represents the best linear fit based on sample observations. Therefore, the regression coefficient for each independent variable measures the change in the dependent variable per unit change in that particular independent variable when all other independent variables are held fixed.

A correlation matrix was computed by the BMD program. Simple correlations provide a measure of association between two variables. The proportion of the variable Y variation which is explained by the variable X is defined by the square of the simple correlation (r^2) for two variables.

Appendix B

Preservice Model: Statistical Results

SAMPLE

The statistical research conducted deals exclusively with active duty Navy enlisted personnel. The data source used was the Navy enlisted master file located in Washington, D.C. There are approximately 460,000 enlisted personnel in the Navy. From the computer files, a random sample of approximately 90,000 enlisted personnel was extracted. A one or a two in the unit's position of an individual's social security number was used as a random selection mechanism. Additional constraints were imposed on the random sample of 90,000 cases in order to generate the samples used in the models. It was required that each case (individual) be male, enlisted, and Regular Navy, and that each case include all information to be used in the analysis.

VARIABLES

The regression model dependent variable was paygrade. The independent variables that were considered to enter the preservice model were Armed Forces Qualifying Test (AFQT), General Classification Test (GCT), Arithmetic Reasoning Test (ARI), Mechanical Comprehension Test (MECH), Clerical Test (CLER), Shop Practices Test (SHOP), marital status, age at entry, black personnel, other minority personnel, and Central Atlantic, Pacific, Southeast, Great Lakes, Southwest, Midwest, Northeast, and Noncontinental United States.

STATISTICAL RESULTS

The independent variables found to be statistically related to paygrade were listed in Table IV-2. The significant probabilities and standard regression coefficients were also listed, as well as their individual contribution to the multiple R^2 value. The correlation matrix for the preservice plus in-service variables evaluated in the models is shown in Table B-1.

SHORTCOMINGS OF THE MODEL

The regional variables are binary variables indicating an individual's region of residence at the time of enlistment. The

Appendix B

TABLE B-1
Preservice Plus In-Service Correlation Matrix

Variables		ED YEARS 1	AFQT 2	GCT 3	ARI 4	MECH 5	CLER 6	SHOP 7	BUSTED 8	BLACK 9	ETHNIC 10
ED YEARS	1	1.0000									
AFQT	2	0.4167	1.0000								
GCT	3	0.5015	0.7467	1.0000							
ARI	4	0.4920	0.7281	0.7461	1.0000						
MECH	5	0.2011	0.6113	0.4426	0.3844	1.0000					
CLER	6	0.2765	0.2583	0.3108	0.3484	0.0983	1.0000				
SHOP	7	0.2163	0.5516	0.4538	0.3830	0.6731	0.0988	1.0000			
BUSTED	8	−0.1118	−0.0894	−0.1002	−0.1074	−0.0590	−0.0275	−0.0534	1.0000		
BLACK	9	−0.0785	−0.3553	−0.3478	−0.3301	−0.3270	−0.1473	−0.3391	0.0619	1.0000	
ETHNIC	10	−0.1513	−0.1344	−0.1311	−0.1462	−0.0700	−0.0422	−0.0402	0.0154	0.1012	1.0000
MARRIED	11	0.2171	0.2363	0.2012	0.2547	0.1351	0.0634	0.1140	−0.0854	−0.0941	−0.1279
CENTRAL	12	0.0012	0.0157	−0.0018	0.0038	0.0118	−0.0081	−0.0119	0.0039	−0.0231	−0.0014
PACIFIC	13	−0.0426	−0.0084	−0.0028	−0.0152	0.0132	0.0355	0.0208	0.0307	−0.0474	0.0102
SOUEAST	14	0.0106	0.0335	0.0446	0.0246	0.0371	−0.0064	0.0399	0.0167	−0.0502	0.0025
GT LAKES	15	0.0025	−0.0140	0.0064	0.0039	0.0101	−0.0077	0.0039	−0.0239	−0.0225	0.0245
MIDWEST	16	0.0218	0.0068	0.0120	−0.0013	−0.0398	0.0270	−0.0134	−0.0327	−0.0368	−0.0019
SOUWEST	17	0.0174	−0.0082	−0.0015	−0.0082	0.0022	0.0182	−0.0040	−0.0099	−0.0069	−0.0250
NONCONUS	18	0.0024	−0.0037	−0.0149	0.0118	0.0174	−0.0231	0.0007	0.0171	0.0338	−0.0560
AGE	19	0.5031	0.1887	0.1981	0.1962	0.0879	0.1028	0.1262	−0.0585	0.0590	−0.0605
RANK	20	0.4562	0.3228	0.5226	0.5749	0.2857	0.1858	0.2697	−0.2469	−0.2418	−0.2318
OPENTRK	21	−0.0121	0.1134	0.0836	0.0495	0.1536	−0.0017	0.1639	−0.0084	−0.0994	0.0014
CLOSEDTK	22	0.0331	0.0472	0.0144	0.0034	0.0791	−0.0081	0.0605	−0.0225	−0.0687	−0.0123
MONTHS	23	0.3208	0.3559	0.3381	0.4036	0.1612	0.0468	0.1377	−0.1014	−0.1840	−0.2125

TABLE B-1—Continued

Variables		MARRIED 11	CENTRAL 12	PACIFIC 13	SOUEAST 14	GT LAKES 15	MIDWEST 16	SOUWEST 17	NONCONUS 18	AGE 19	RANK 20
MARRIED	11	1.0000									
CENTRAL	12	0.0080	1.0000								
PACIFIC	13	−0.0567	−0.0277	1.0000							
SOUEAST	14	−0.0675	−0.0228	−0.0486	1.0000						
GT LAKES	15	0.0188	−0.0524	−0.1073	−0.0881	1.0000					
MIDWEST	16	−0.0260	−0.0800	−0.1637	−0.1345	−0.3097	1.0000				
SOUWEST	17	0.0608	−0.0147	−0.0301	−0.0248	−0.0571	−0.0871	1.0000			
NONCONUS	18	0.0881	−0.1412	−0.0843	−0.0693	−0.1596	−0.2437	−0.0449	1.0000		
AGE	19	0.1931	−0.0681	−0.0401	0.0048	0.0189	−0.0389	0.0186	0.0316	1.0000	
RANK	20	0.5175	0.0349	−0.0268	−0.0112	−0.0157	0.0126	0.0070	0.0853	0.1773	1.0000
OPENTRK	21	−0.0138	0.0134	−0.0118	−0.0178	0.0186	−0.0087	0.0191	0.0163	−0.0224	0.0655
CLOSEDTK	22	0.0508	−0.0141	0.0279	0.0073	0.0148	−0.0074	−0.0005	−0.0070	−0.0007	−0.0482
MONTHS	23	0.5775	0.0570	−0.0342	−0.0190	−0.0114	0.0056	0.0078	0.0857	0.1125	0.8329

		OPENTRK 21	CLOSEDTK 22	MONTHS 23
OPENTRK	21	1.0000		
CLOSEDTK	22	−0.2439	1.0000	
MONTHS	23	−0.0123	−0.0092	1.0000

Appendix B

random sample of 90,000 cases (individuals) extracted from the master enlisted files listed the individuals in ascending order of social security number. The first three numbers of an individual's social security number indicate the region in which the social security card was issued to that individual.

Five thousand cases from the possible 90,000 provided by the Bureau of Naval Personnel (BUPERS) were sampled. If an individual was male, active duty, and had data on the variables included in the model, that individual met the constraints imposed and was therefore eligible to be included in the sample of 5,000. The cases were evaluated one by one to determine whether or not the above constraints were met. The computer program used to extract the 5,000 cases began evaluating cases at the beginning of the BUPERS tape reel and worked through approximately 70,000 cases before the 5,000 cases for the sample were extracted. Because the reel did not run through the entire 90,000 cases, individuals with the highest social security numbers were not considered.

As a result of the above method of sample extraction, a random distribution of regions was not obtained. That is, the sample regional distribution was not an accurate distribution of the Navy enlisted population. In many cases, the region in which an individual was issued a social security number was the same as that individual's region of residence at the time of enlistment. When this was the case, and when the social security numbers for the region were very high, those individuals from the region were not considered.

The above regional bias was mitigated, in part, by the fact that often the region in which an individual was issued a social security number was not that individual's region of residence at enlistment. The result of this fact was that each regional variable was well enough represented in the sample to be considered by the model. Therefore, investigation of the regional variables was not adversely affected. Care was taken, however, in interpreting findings with respect to the regional variables.

A second bias of the model was the distribution of paygrade. The most pervasive constraint imposed on each case included in the sample required that information be available for each variable considered in the model for that case. Because of the newness of the computerized enlisted records system, as well as the method of filing new data in the enlisted files, an individual who recently entered the Navy was more likely to have a complete record than a more senior enlisted man. For this reason, the constraint requiring that all information utilized in the model be present excluded disproportionately more senior enlisted personnel than junior enlisted personnel. As a result, the distribution of rank in the sample did not represent the true frequency distribution of rank in the population of Navy enlisted personnel.

The breakdown by rank of personnel included in the sample of 5,000 was as follows: E-1, 129; E-2, 1,501; E-3, 759; E-4, 660; E-5, 1,529; E-6, 418; and E-7, 4. As can be seen, paygrades E-1 through E-6 were well represented in the sample. Paygrade E-7 was poorly represented, and paygrades E-8 and E-9 were not represented at all. Again, it must be said that the sample was not representative of the Navy enlisted population. Because of the large number of cases in each of paygrades E-1 through E-6, however, the model was not adversely affected by the lack of paygrade randomness when considering these paygrades. But nothing can be said about the significant preservice variables which define paygrade for paygrades E-7 through E-9, nor the impact of these variables on minorities in paygrades E-7 through E-9.

The frequency distribution of the time-in-service variable clearly indicated that the probability of an individual being excluded from the sample was greatly increased by seniority. The mean value for time in the Navy was thirty-five months. The median value was twenty-one months. Although the range of time in the Navy for sample personnel was two months to fifteen years, approximately 70 percent of the members of the sample had less than or equal to four years active duty as of June 1975. Therefore, the distribution of time in the Navy for personnel in the sample was skewed to the left. Because of the greater number of people in the Navy for periods of time two years and less, the reliability of the model was greater for this period of time.

Appendix C

*Preservice Plus In-Service Model:
Statistical Results*

VARIABLES

The regression model's dependent variable remained paygrade. The independent variables that were considered for the preservice model remained as candidates in this model. In addition, occupations with open advancement potential, occupations with closed advancement potential, discipline record, and time in service were considered for the model.

STATISTICAL RESULTS

The independent variables found to be statistically related to paygrade were listed in Table IV-3. The significant probabilities and standard regression coefficients were also listed, as well as their individual contribution to the multiple R^2 value.

SHORTCOMINGS OF THE MODEL

The sample for this model was the same as that of the preservice regression model. Because the sample was the same, the sample bias was also the same. First, the regional frequency distribution of the sample did not accurately represent the regional frequency distribution of the Navy enlisted population. Because all regions were well enough represented in the sample, however, inferences were drawn concerning the relative relationships of the regional variables with paygrade level attainment. Second, the sample paygrade frequency distribution did not reflect the true paygrade distribution of the enlisted population. Again, because only paygrades E-1 through E-6 were well represented in the sample, conclusions were drawn concerning the impact of the explanatory variables on personnel promotion opportunities in these paygrades only. Finally, the sample time-in-service distribution did not reflect the true population distribution of time in the Navy.

The Navy has time-in-paygrade requirements that must be fulfilled before an individual is eligible for promotion consideration. This fact distorted somewhat the results of the model. This was a bias of the months variable. Because of the short mean length of time in the Navy and the skewness of the rank distribution towards the lower paygrades for the sample, how-

Appendix C

ever, the distortion was not considered severe. One reason for this was that, although the time-in-paygrade requirements were applicable for all paygrades, they were more restrictive for paygrades E-6 through E-9. As already stated, paygrades E-7 through E-9 were poorly represented in the sample. There were two additional reasons why the time-in-grade constraint was mitigated. First, the time-in-grade requirements were lengthened to the current requirements only recently. That is, the time-in-grade requirements were gradually increased since the end of the United States' involvement in Vietnam. Therefore, during the war, the time-in-grade requirements were even less restrictive than today. Second, a percentage of recruits graduating from formal school training are automatically advanced to E-4 upon graduation. The standard time-in-grade and time-in-service requirements do not apply in these cases.

An additional bias was imposed by the fact that there is an "up-or-out" policy in the Navy. Again, the distortion was not considered severe. The up-or-out policy pertains only to those individuals wishing to reenlist after the expiration of their initial or succeeding service obligation contracts. Normally, initial enlisted obligated service is for four years. Greater than 70 percent of the sample had less than or equal to four years' active duty. Furthermore, most of the so-called up-or-out emphasis is placed on being in an occupation which is not overmanned rather than on requiring upward advancement.

Appendix D

*Separate Black and
Other Minority Personnel Models:
Statistical Results*

SAMPLE AND VARIABLES

Separate models for both black and other minority members were derived from the sample of 5,000. The black subsample consisted of the 524 blacks in the sample. The other minority subsample consisted of the 390 other minority members in the sample. The dependent variable remained paygrade for both models. The explanatory variables that were considered for the preservice plus in-service model remained as candidates in these models with the exception of the race-ethnic variables.

It is noted that the aggregate sample consisted of approximately 18 percent minority personnel. Therefore, a comparison between the minority models and the aggregate model was not a direct comparison of minorities' and nonminorities' advancement functions. Reliable inferences, however, could be made from this section of the study when comparing minorities' and nonminorities' advancement functions.

Because the samples for these two models were subsamples of the original 5,000 case sample, the nature of the bias was the same. That is, the subsample frequency distributions of paygrade, time in the Navy, and the regional variables did not reflect the true frequency distributions of these variables in the enlisted population. In the earlier models, conclusions had to be made with caution. Additional caution had to be exercised in analyzing the results of the black and other minority models.

In the sample of 5,000, paygrades E-1 through E-6 were well represented. Therefore, reliable conclusions were drawn concerning the impact of the explanatory variables on the dependent variable for personnel in these paygrades. The breakdown by paygrade of personnel included in the black subsample was as follows: E-1, 33; E-2, 308; E-3, 76; E-4, 42; E-5, 60; and E-6, 5. The breakdown for personnel included in the other minority subsample was as follows: E-1, 16; E-2, 266; E-3, 28; E-4, 59; E-5, 19; and E-6, 2.

The frequency distribution of paygrade for both the black subsample and the other minority subsample was more skewed to the left than the aggregate sample paygrade frequency distribution. The same holds true for the time-in-service frequency distribution. That is, there were paygrades, particularly the higher ones, in both the black subsample and the other minority sub-

Appendix D 223

sample which were poorly represented. Thus, the reliability of the models derived from the above subsamples was decreased. This was more true of the other minorities model than for the black model.

STATISTICAL RESULTS

The regression equation and the appropriate statistics for the black model are shown in Table D-1. The results of the regression equation for the other minorities model are shown in Table D-2.

TABLE D-1
Black Model Statistics: Significant Variables Listed
in Order of Significance

Variable	Standard Regression Coefficient	Contribution to R^2
Time in Navy	0.663	0.5850
Discipline record	—0.238	0.0722
AFQT	0.129	0.0526
Years of education	0.122	0.0201
SHOP	0.079	0.0086
Marital status	0.059	0.0031
GCT [a]	0.073	0.0029
Open occupations [a]	—0.052	0.0025
		0.7470

[a] Variable is significant at the .025 level of significance. All other variables are significant at the .005 level of significance.

TABLE D-2
Other Minority Model Statistics: Significant Variables
Listed in Order of Significance

Variable	Standard Regression Coefficient	Contribution to R^2
Time in Navy	0.500	0.3344
AFQT	0.257	0.1291
Discipline record	—0.188	0.0374
Years of education	0.111	0.0114
Closed occupations	—0.106	0.0105
ARI [a]	0.107	0.0061
		0.5289

[a] Variable is significant at the .05 level of significance. All other variables are significant at the .005 level of significance.

Appendix E

Cross-Sectional Model: Statistical Results

SAMPLE AND VARIABLES

Individuals selected for the cross-sectional sample were screened in the same manner, with one exception, as those individuals included in the original preservice plus in-service sample. The exception to this was that the individuals included in the cross-sectional sample entered the Navy in either April, May, June, or July of 1971. Therefore, this model constructed an advancement function for individuals who had approximately four years of active Navy service as of the file date.

The variables studied in this model were the same as those studied in the preservice plus in-service model. Time in service was still considered in order to accommodate the differences of up to three months. The relative significance of this variable, however, was expected to be much less. The multiple correlation, therefore, was expected to drop considerably.

The bias accounted for by time in the Navy in the earlier models was no longer with us. Furthermore, because all 90,000 cases of the random file tape were screened to find this cross-sectional sample of 2,204 individuals, the regional bias was eliminated. Paygrade bias remained, however. The sample paygrade was as follows: E-1, 4; E-2, 33; E-3, 233; E-4, 954; E-5, 978; and E-6, 2. Again, the problem of poor representation in several paygrades is evident. The reliability of the model was reduced for all enlisted paygrades other than E-3, E-4, and E-5.

STATISTICAL RESULTS

The regression equation and the appropriate statistics for the cross-sectional model are shown in Table E-1.

Appendix E

TABLE E-1
Cross-Sectional Model Statistics: Significant Variables Listed in Order of Significance

Variable	Standard Regression Coefficient	Contribution to R^2
Discipline record	—0.369	0.1225
Time in paygrade	—0.336	0.0931
GCT	0.146	0.1070
Marital status	0.130	0.0210
ARI	0.130	0.0178
SHOP	0.068	0.0054
Time in Navy	0.073	0.0050
Years of education	0.075	0.0051
Black personnel	0.059	0.0025
Open occupations	0.048	0.0023
AFQT [a]	0.064	0.0018
CLER [a]	0.041	0.0015
		0.3850

[a] Variable is significant at the 0.025 level of significance. All other variables are significant at the 0.005 level of significance.

//
Appendix F

*Performance Evaluation Model:
Statistical Results*

SAMPLE AND VARIABLES

Only the performance evaluations for E-5 and above are filed on tape. Therefore, individuals selected for the sample were in paygrades E-5 and above. The regression model's dependent variable remains paygrade. The independent variables that were considered for the preservice plus in-service model remained as candidates in this model. In addition, an individual's performance, appearance, cooperativeness, reliability, conduct, resourcefulness, leadership, and overall and equal opportunity evaluations (member's ability to deal with individuals of all race-ethnic groups in a nondiscriminatory manner) were also considered for the model.

The regional variable bias no longer was in the sample. This was true because all 90,000 cases of the random file were screened to generate the sample of 3,109 individuals. The paygrade bias remained, however. The enlisted paygrades were represented as follows: E-5, 1,959; E-6, 1,139; E-7, 11. Only paygrades E-5 and E-6 are well represented. E-7 is poorly represented, and paygrades E-8 and E-9 are not represented at all. The model was reliable for describing the advancement function only for paygrades E-5 and E-6.

STATISTICAL RESULTS

The regression equation and the appropriate statistics for the performance evaluation model are shown in Table F-1. The correlation matrix for the preservice plus in-service variables evaluated in the performance model is shown in Table F-2.

Appendix F

TABLE F-1
Performance Evaluation Model Statistics: Significant Variables Listed in Order of Significance

Variables	Standard Regression Coefficient	Contribution to R^2
Time in Navy	0.560	0.2594
Time in paygrade	—0.280	0.0881
ARI	0.131	0.0474
Open occupations	0.115	0.0229
Leadership evaluation	0.112	0.0227
SHOP	0.091	0.0098
Central Atlantic region	0.098	0.0075
Appearance	0.092	0.0071
Great Lakes region	—0.069	0.0046
Closed occupations	—0.064	0.0044
Southeast region	—0.068	0.0036
GCT	0.072	0.0029
		0.4804

Note: All variables are significant at the 0.005 level of significance.

TABLE F-2
Performance Evaluation Model Correlation Matrix

		ED YEARS 1	AFQT 2	GCT 3	ARI 4	MECH 5	CLER 6	SHOP 7	PERFORM 8	APPEAR 9	COOPER 10
ED YEARS	1	1.0000									
AFQT	2	0.2271	1.0000								
GCT	3	0.3503	0.5758	1.0000							
ARI	4	0.3541	0.5459	0.6274	1.0000						
MECH	5	0.0970	0.5408	0.3088	0.2261	1.0000					
CLER	6	0.1618	0.1154	0.1938	0.2546	0.0100	1.0000				
SHOP	7	0.0898	0.4075	0.2647	0.1643	0.4182	−0.0738	1.0000			
PERFORM	8	0.0623	0.0562	0.0642	0.0920	0.0384	0.1167	−0.0185	1.0000		
APPEAR	9	0.0432	−0.0073	0.0058	0.0136	−0.0247	0.0604	−0.0346	0.3919	1.0000	
COOPER	10	0.0738	0.0409	0.0379	0.0547	0.0059	0.0871	−0.0351	0.7211	0.4293	1.0000
RELIAB	11	0.0623	0.0568	0.0450	0.0744	0.0898	0.1031	−0.0088	0.8404	0.4364	0.7556
CONDUCT	12	0.0508	−0.0104	0.0107	0.0406	−0.0066	0.0543	−0.0472	0.4848	0.5393	0.5535
RESOURC	13	0.0609	0.0486	0.0410	0.0783	0.0244	0.0954	−0.0415	0.8057	0.4049	0.7768
LEADDIR	14	0.0302	0.0419	0.0320	0.0588	0.0387	0.0701	−0.0401	0.7758	0.4502	0.7627
LEADCOUN	15	0.0325	0.0212	0.0059	0.0515	−0.0039	0.0523	−0.0614	0.7039	0.4626	0.7679
OVERALL	16	0.0544	0.0413	0.0478	0.0714	0.0182	0.0905	−0.0365	0.8508	0.5080	0.8405
BEHAVIOR	17	−0.0178	−0.0010	−0.0186	0.0033	0.0144	0.0009	0.0227	−0.1057	−0.1053	−0.1201
EQUALOP	18	0.0363	−0.0176	0.0062	0.0217	−0.0426	0.0749	−0.0671	0.5675	0.3900	0.6724
BUSTED	19	0.0047	0.0158	−0.0216	0.0003	0.0204	0.0009	0.0125	−0.0170	−0.0020	−0.0099
BLACK	20	−0.0322	−0.3102	−0.2412	−0.2482	−0.2114	−0.0130	−0.0671	−0.0723	0.0394	−0.0431
ETHNIC	21	−0.0097	−0.0883	−0.0542	−0.0396	−0.0468	−0.1007	−0.1303	−0.0102	−0.0203	0.0133
MARRIED	22	−0.0434	0.0206	0.0634	−0.0068	0.0608	−0.0086	−0.0500	0.0556	0.0851	0.0852
CENTRAL	23	0.0306	0.0353	0.0068	0.0051	0.0234	−0.0401	0.0049	0.0288	0.0161	0.0565
PACIFIC	24	−0.0180	0.0147	0.0086	0.0003	0.0241	0.0027	0.0313	0.0156	0.0026	0.0014
SOUEAST	25	0.0031	−0.0022	−0.0388	−0.0243	0.0047	0.0045	0.0070	−0.0318	0.0113	−0.0151
GT LAKES	26	−0.0119	−0.0127	−0.0081	−0.0023	0.0054	−0.0611	−0.0077	0.0142	0.0396	0.0094
MIDWEST	27	−0.0053	−0.0287	−0.0016	−0.0194	−0.0547	−0.0122	−0.0135	−0.0016	−0.0226	−0.0401
SOUWEST	28	−0.0257	−0.0538	−0.0440	−0.0277	−0.0028	−0.0419	−0.0186	−0.0039	0.0116	0.0113
NONCONUS	29	−0.0141	−0.0249	−0.0179	−0.0167	0.0003	−0.0439	−0.0097	−0.0219	−0.0213	−0.0251
AGE	30	0.4987	0.1277	0.1504	0.1527	0.1227	0.0297	−0.0128	0.0403	0.0358	0.0585
RANK	31	0.0781	0.2107	0.1585	0.2199	0.1922	0.0946	−0.0891	0.2029	0.1987	0.2176
OPENTRK	32	−0.0279	0.1019	0.0969	0.0549	0.1484	−0.0126	0.0885	−0.0287	0.0234	−0.0298
CLOSEDTK	33	−0.0224	−0.0027	−0.0863	−0.0655	0.0332	−0.0219	0.1139	0.0261	0.0077	0.0183
MONTHS	34	−0.0833	0.0041	−0.0865	0.0269	0.0379	−0.1658	−0.1751	0.0956	0.0552	0.0908

Appendix F

TABLE F-2—Continued

		RELIAB 11	CONDUCT 12	RESOURC 13	LEADDIR 14	LEADCOUN 15	OVERALL 16	BEHAVIOR 17	EQUALOP 18	BUSTED 19	BLACK 20
RELIAB	11	1.0000									
CONDUCT	12	0.5256	1.0000								
RESOURC	13	0.8126	0.5044	1.0000							
LEADDIR	14	0.7989	0.5227	0.7697	1.0000						
LEADCOUN	15	0.7381	0.5788	0.7452	0.8252	1.0000					
OVERALL	16	0.8658	0.5983	0.8591	0.8640	0.8291	1.0000				
BEHAVIOR	17	−0.1360	−0.2434	−0.1105	−0.1363	−0.1527	−0.1469	1.0000			
EQUALOP	18	0.5917	0.4596	0.6174	0.6096	0.6610	0.6920	−0.0804	1.0000		
BUSTED	19	−0.0137	−0.0114	−0.0107	−0.0036	−0.0181	−0.0099	−0.0034	0.0027	1.0000	
BLACK	20	−0.0652	−0.0497	−0.0457	−0.0607	−0.0810	−0.0526	0.0164	0.0643	−0.0058	1.0000
ETHNIC	21	−0.0202	−0.0073	−0.0026	−0.0019	0.0065	−0.0062	0.0068	0.0163	−0.0035	−0.0264
MARRIED	22	0.0754	0.0742	0.0734	0.1153	0.1192	0.1024	−0.0448	0.0606	0.0177	−0.0081
CENTRAL	23	0.0513	0.0357	0.0381	0.0622	0.0520	0.0560	−0.0239	0.0442	−0.0054	−0.0301
PACIFIC	24	0.0093	0.0132	0.0047	0.0119	−0.0192	0.0015	−0.0117	0.0129	0.0638	−0.0048
SOUEAST	25	−0.0320	−0.0130	−0.0068	−0.0213	0.0182	−0.0087	−0.0073	0.0069	−0.0097	0.0320
GT LAKES	26	0.0303	0.0091	0.0164	0.0443	0.0197	0.0271	0.0026	0.0107	−0.0147	0.0298
MIDWEST	27	−0.0005	−0.0268	−0.0122	−0.0411	−0.0809	−0.0282	−0.0012	−0.0146	−0.0143	−0.0215
SOUWEST	28	−0.0034	0.0210	−0.0026	0.0093	0.0094	0.0014	−0.0163	−0.0167	−0.0065	0.0467
NONCONUS	29	−0.0331	−0.0107	−0.0291	−0.0559	−0.0457	−0.0305	0.0466	−0.0160	−0.0090	0.0121
AGE	30	0.0549	0.0517	0.0472	0.0478	0.0662	0.0465	−0.0166	0.0445	0.0165	0.0006
RANK	31	0.2256	0.1682	0.2079	0.2573	0.2628	0.2359	−0.0322	0.1708	−0.0215	−0.0900
OPENTRK	32	−0.0287	0.0058	−0.0385	0.0014	−0.0200	−0.0356	0.0240	−0.0160	−0.0251	−0.0716
CLOSEDTK	33	0.0400	0.0132	0.0250	0.0384	0.0375	0.0295	−0.0119	0.0248	−0.0076	0.0032
MONTHS	34	0.1022	0.0723	0.1070	0.1605	0.1679	0.1132	−0.0426	0.0762	0.0096	−0.0111

TABLE F-2—Continued

		ETHNIC 21	MARRIED 22	CENTRAL 23	PACIFIC 24	SOUEAST 25	GT LAKES 26	MIDWEST 27	SOUWEST 28	NONCONUS 29	AGE 30
ETHNIC	21	1.0000									
MARRIED	22	0.0303	1.0000								
CENTRAL	23	0.0367	0.0174	1.0000							
PACIFIC	24	−0.0214	−0.0234	−0.0357	1.0000						
SOUEAST	25	0.0419	0.0216	−0.0420	−0.0458	1.0000					
GT LAKES	26	−0.0295	0.0654	−0.0904	−0.0986	−0.1161	1.0000				
MIDWEST	27	−0.0087	−0.0570	−0.1079	−0.1177	−0.1388	−0.2984	1.0000			
SOUWEST	28	−0.0013	−0.0249	−0.0406	−0.0443	−0.0521	−0.1122	−0.1840	1.0000		
NONCONUS	29	−0.0072	0.0288	−0.0641	−0.0700	−0.0824	−0.1773	−0.2118	−0.0797	1.0000	
AGE	30	−0.0091	0.0382	0.0137	−0.0148	0.0424	−0.0160	−0.0745	−0.0148	0.0588	1.0000
RANK	31	−0.0265	0.1806	0.1477	−0.0175	0.1180	0.1117	−0.0953	−0.0093	−0.0852	0.0489
OPENTRK	32	0.0231	−0.0608	−0.0145	−0.0181	−0.0129	0.0619	−0.0291	−0.0101	0.0160	−0.0308
CLOSEDTK	33	0.0117	0.0556	0.0140	0.0162	0.0074	−0.0343	0.0144	0.0316	0.0229	−0.0179
MONTHS	34	0.0052	0.2580	0.0949	−0.0463	0.0807	0.0539	−0.0208	0.0015	−0.0735	−0.0235

		RANK 31	OPENTRK 32	CLOSEDTK 33	MONTHS 34
RANK	31	1.0000			
OPENTRK	32	0.1210	1.0000		
CLOSEDTK	33	−0.0840	−0.2324	1.0000	
MONTHS	34	0.5093	−0.0933	0.0734	1.0000

Index

ACT, 47
AD. *See* Attention to Detail
ADCOP. *See* Associate Degree Completion Program
advancement examinations. *See* advancement system
advancement function, 72
 black model, 76, 86-88, 221-23
 cross-sectional model, 76, 88-91, 171, 225-27
 other minorities model, 76, 86-88, 221-23
 performance evaluation model, 76, 91-94, 171, 229-34
 preservice model, 76-81, 211-16
 preservice plus in-service model, 81-86, 171, 217-19
 variables in, 73-75
advancement system, 134-65, 204-05
 alternative testing methods, 145-46, 196
 alternative weighting of criteria, 147-48, 195-96
 criteria for enlisted, 136-42
 effect of manpower pool on, 144
 eligibility criteria, 193-96
 "essential subjects" test, 139
 examination cut-off scores, 143-45
 examinations, 136, 137-39
  role in, 141-45, 147, 194-96
 face-to-face counseling, 197
 procedure for enlisted, 136-42
 procedure for officers, 159-65
 role of career pattern in, 161, 162-63, 164, 205
 role of fitness reports in, 161, 163-65, 205
 role of military justice in, 148-51, 198-99
 role of performance evaluations in, 136, 137, 140-41, 147, 197-98
 role of seniority in, 136
 role of warfare speciality in, 161, 165, 205
 time-in-grade requirements, 137, 162, 163, 218-19
 "whole-man" concept, 137, 179, 193

affirmative action, 1, 154-55. *See also* Command Affirmative Action Plans
 defined, 6
Affirmative Action Plan Development Workshop, 28
Affirmative Action Plan Revision Track, 28
AFQT. *See* Armed Forces Qualifying Test
Air Force, 60, 65, 129n
Albany (New York), 41
alternative training methods, 192
American Society for Personnel Administration, 46n
AOCS. *See* Aviation Officer Candidate School
Applicant Qualification Test (AQT), 131
Arceneaux, E. E., 27n
Arithmetic Reasoning (AR), 111, 112
Arithmetic Reasoning Test (ARI), 73-93 passim, 111-12, 115, 123, 212, 213, 214, 223, 227, 231, 232
Arlington (Virginia), 16n
Armed Forces Examining and Entrance Station, 112
Armed Forces Qualifying Test (AFQT), 49, 50, 51, 52, 53, 60, 61, 62, 73-93 passim, 100, 110, 112-13, 212, 213, 214, 223, 227, 232
Armed Services Vocational Aptitude Battery (ASVAB), 49, 50, 51, 76n, 110-15, 116, 125
Army, 60, 65, 129n
Asian American, 74, 76n
Associate Degree Completion Program (ADCOP), 127
ASVAB. *See* Armed Services Vocational Aptitude Battery
Atlanta (Georgia), 16n
Attention to Detail (AD), 111
Auerbach, Herbert A., 120n, 146n
Aviation Officer Candidate School (AOCS), 65

Bascom, Lionel, 60n, 184n

235

Basic Test Battery (BTB), 73-93 passim, 110-12, 115, 116, 125, 170, 176, 177, 178, 179
Battelle, R. B., 62n
Bergman, Brian A., 53n, 80n, 105n, 121n, 146n
Bilinski, Chester R., 119
Binkin, Martin, 52n
Biomedical Computer Programs, 208
black model. *See* advancement function, black model
blacks. *See also* minorities
  in the Marine Corps, 2-5
  in the Naval Academy, 65
  in the Navy, 2-5
Bloom, G. F., 6n
Broadened Opportunity for Officer Selection and Training (BOOST), 157-58
Bureau of the Census, 2n, 57n, 63n, 64
Bureau of Labor Statistics, 59
Bureau of National Affairs, 46n
Bureau of Naval Medicine, 50
Bureau of Naval Personnel, 4, 5, 20, 22n, 23n, 24n, 41, 54, 59, 75, 82n, 86n, 97, 106n, 110n, 111, 113n, 114n, 115n, 128n, 130, 132n, 135, 136n, 138, 142n, 152, 157n, 160, 161n, 163, 164, 167n, 168n, 169n, 174, 175, 177n, 180n, 182, 215
  chief of, 22n, 23, 37, 144
Bureau of Naval Reactors, 132

Campion, J. E., 146n
Camp Pendleton, 34, 155n
career,
  counseling, 155, 158-59, 199-200
  defined, 166-67
Career Recruiter Force (CRF), 44-45
Career Reenlistment Objectives (CREO), 75, 82n, 169-71, 177
  and lateral transfer, 151-54
Cassell, Frank H., 46
Chicanos, 47
Chinese, 74, 76n
Civil Rights Act of 1964, as amended in 1972, 1, 26
Civil War, 9-10
Class A school. *See* formal school training
Class C school. *See* formal school training
Cleary, T. Anne, 117n

Clerical Test (CLER), 73-93 passim, 111-12, 212, 213, 214, 227, 232
closed rating, 75, 82, 85-86, 87, 151-54 passim, 169-70, 213, 214, 224, 232, 233, 234. *See also* Career Reenlistment Objectives
Coffey, Kenneth J., 59
Columbus (Ohio), 41
Combat Arms Enlistment Bonus, 103-04
Command Affirmative Action Plans, 23, 24, 25-26, 28
Command Information Team, 30
Command Training Team, 29
Congress, 51, 57, 66, 126, 127, 167, 190, 191
Continental navy, 8
counselor rating, 45
Counter Racism/Equal Opportunity Workshop, 28, 29
Coursey, John J., 149n
Cowin, Ronald M., 118n
Crawford, Kenneth S., 150n
CREO. *See* Career Reenlistment Objectives
CRF. *See* Career Recruiter Force
cross-sectional model. *See* advancement function
Cuban American, 74, 76n
Cultural Expression in the Navy Workshop, 29, 30
Curran, Thomas E., 123n
cut-off scores, 143-45, 187, 193-96

Dalfiume, Richard M., 9n
Dallas (Texas), 41
Daniel, Johnnie, 149n
Defense Appropriations Bills, 51, 52
Defense Manpower Commission, 100n, 113n, 167, 169n
Defense Manpower Data Center, 3, 96n
Defense Race Relations Institute (DRRI), 22, 30
Defense Research Projects Agency, 99n
Department of Defense, 22, 30, 32, 59, 60, 72n, 96n, 137, 140, 149, 174, 181n, 191
  deputy assistant secretary, 3, 96n, 148n, 149n
  Directive 1100.15, 1, 6
  occupational group, 96, 97, 128
Department of Health, Education and Welfare (HEW), 125, 190

Index

disadvantaged,
 defined, 6n
Dixon, W. J. 208
Docter, Richard, 120n
Doré, Russell, 107n, 108n
DRRI. *See* Defense Race Relations Institute
Duffy, Thomas M., 123n, 124n

educational opportunities,
 in the military, 155-58
EEOC. *See* Equal Employment Opportunity Commission
Electronics Information (EI), 111
Electronics Technician Selection Test (ETST), 75, 76n, 111-12
Emancipation Proclamation, 9-10
encounter groups, 23
enlisted advancement. *See* advancement system
Enlisted Navy Recruiter Orientation Schools (ENRO), 41-43
enlisted performance evaluations. *See* advancement system, role of performance evaluations in
Enlistment Bonus Program. *See* Marine Corps, Enlistment Bonus Program
enlistment contract, 102-05
 length of, 104
enlistment options, 102-04
ENRO. *See* Enlisted Navy Recruiter Orientation Schools
EOPS. *See* Equal Opportunity Program Specialists
EOQI. *See* Equal Opportunity Quality Indicators
Equal Employment Opportunity Commission (EEOC), 117
equal opportunity,
 defined, 6
 programs, 162. *See also* race relations programs
Equal Opportunity Program Specialists (EOPS), 27-30
Equal Opportunity Quality Indicators (EOQI), 28, 30
Eskimo, 74, 76n
"essential subjects" test. *See* advancement system, "essential subjects" test
ETST. *See* Electronics Technician Selection Test
examination cut-off scores. *See* advancement system, examination cut-off scores; cut-off scores

Executive Order 9981, 12-13
Executive Order 11246, 1, 5, 6, 13-14, 26
Executive Seminars, 23, 24, 25, 26, 33. *See also* Navy Race Relations Program

Fahy, Charles H., 13
FAR. *See* Flight Aptitude Rating
Filipino, 74, 76n
Flag Seminars, 23, 24. *See also* Navy Race Relations Program
Fletcher, John D., 123n
Flight Aptitude Rating (FAR), 131
flight training, 65, 127n, 129-30, 131-32, 204
formal school training, 99n, 100n, 102-05, 153, 175, 176, 177, 178, 179, 187-88, 189, 201
 alternative selection methods, 119-22
 alternative training methods, 122-23
 entrance requirements, 104-05, 110-15
 and experimental enrollment of non-eligibles, 118-19
 inflated mental aptitude criteria, 116-17
 in-fleet assignment, 104-05
 and minorities, 114, 23
Forrestal, James, 12, 13
"4F," 48
France, 9
Francis, H. Minton, 60

Garcia, John, 53n
General Classification Test (GCT), 73-93 passim, 111-12, 115, 123, 212, 213, 214, 223, 227, 231, 232
General Educational Development (GED), 51, 57, 127, 155-56, 157
General Electric Co., 39
General Science (GS), 111
General Services Administration, 46
Ghiselli, Edwin E., 117n
Great Lakes (Illinois), 41
GS. *See* General Science
Guaranteed Assignment Retention Detailing (GUARD), 179-80
guaranteed enlistment contract, 102-03, 105

Harrison, George Brooks, 9n, 10n
HEW. *See* Department of Health, Education and Welfare

Hinsvark, Don G., 71n
Holmen, Milton G., 120n
House Committee on Armed Services, 17n, 21
Human Relations Council, 20
Human Relations Instructors (HRI), 30
human resource management, 7
Human Resource Management Centers, 28
Human Resource Management Program, 32
Human Resources Development Project Office, 22

Indians, 6n, 74, 76n
institutional racism, 24, 25, 26-27

James, Jim, 143n
Japanese, 74, 76n
Jehn, Christopher, 143n
Job Oriented Basic Skills (JOBS), 125-26, 190
job-sample test, 146, 196. *See also* advancement system, alternative testing methods
Johnson, Louis B., 13
Johnson, Lyndon B., 13
Junior Reserve Officer Training Corps (JROTC), 15, 17

Kaylor, Robert, 34n
Korean, 74, 76n
Korean War, 13

Lambert, Joseph, 53n, 80n, 105n, 121n, 146n
Landman, R. J., Sr., 149n
lateral transfer, 151-54, 201-02
Leahy, Wm. Rick, 121n
Lent, Richard H., 120n, 146n
Levin, Lowell S., 120n, 146n
Lincoln, Abraham, 9
Lindsey, Robert, 34n
line-staff relationship, 37, 39-40
Lockman, Robert F., 122n
Los Angeles, 17

McNamara, Robert S., 61
Macon (Georgia), 41
Maintenance Track, 28, 29-30
Malayans, 6n
Manila, 10
manpower planning programs, 154-55. *See also* Career Reenlistment Objectives
Marine Corps,
 affirmative action in, 16-17

commandant, 14n, 16, 30n, 31n, 38
Enlistment Bonus Program, 103
equal opportunity in, 12-17, 30-33
equal opportunity policy, 1-2, 6, 14, 16
Human Relations Instructors Institute, 30, 31n
Human Relations Program, 30-33
 Cycle I, 31, 32
 Cycle II, 32
 Cycle III, 32
Manpower Planning, Programming and Budgeting, 4, 5, 55, 97, 135, 160
Recruiting Command, 163n
 assistant chief of staff, 38
 deputy chief of staff for manpower, 38
 director, personnel procurement division, 38
recruiting organization, 36-40, 186
Marine Detachment Reaction Force, 18
Mathematics Knowledge (MK), 111
Meacham, Merle, 107n, 108n
Mechanical Comprehension (MC), 111
Mechanical Comprehension Test (MECH), 76n, 111-12, 212, 213, 214, 232
mental groups, 21, 49, 50, 53-56, 62
mental testing, 48-53, 99, 100-102, 110-19, 187-88, 189
messman branch, 11
Mexican War, 9
Mexican American, 74, 76n
Middle Management Actions to Counter Racism Workshop, 29
military justice, 148-51, 198-99
 and command's organizational effectiveness, 150-51, 198-99
military occupational specialty. *See* advancement system, role of warfare specialty in
Military Rights and Responsibilities Workshop, 29, 30
Millington (Tennessee), 22
minorities,
 advancement of,
  and cut-off scores, 143-45
 availability of, 185-86
  in the professions, 63-65, 129
 and career eligibility, 183-84
 enlistment of,
  reduction in, 58-60

Index

and formal school eligibility requirements, 113-15
and lateral transfer, 153-54
and manpower planning programs, 154-55
and military justice, 148-51, 198-99
and Naval Academy, 132-33
occupational distribution of, 96-98, 127-33
and officer retention, 181-83
paygrade distribution, 134-36, 159-61
penetrating communities of, 41, 46-47
performance evaluation's effect on reenlistment eligibility of, 173-74
promotion rates of, 72-94
recruiting of, 36, 39-40, 42-43, 45, 46-48, 63
  economy's effect on, 57-58, 59
  service standards' effect on, 53-60
reenlistment eligibility of, 173-74
retention of, 171-74
upgrading of, 115-27, 155-59, 178-79, 186, 188-92
  cost-effectiveness of, 116-17, 119, 126-27
statistical analysis of, 72-94
Mitchell, Howard E., 35n
MK. *See* Mathematics Knowledge
mobility,
  defined, 6
Mongolians, 6n
Montford Point (North Carolina), 12
Morgan, Fred, 99n, 100n, 101
multiple stepwise regression model, 207-10

NAAP. *See* Navy Affirmative Action Plan of 1976
Naval Academy, 65, 66, 68, 69, 70, 71, 132, 133, 157, 158, 182, 203, 204
Naval Examining Center (NAVEXAMCEN), 147
Naval Flight Officer Program, 131
naval flight officers (NFO), 181
Naval Operations,
  chief of, 1n, 2n, 14n, 15, 21n, 23n, 26n, 33, 34n, 37, 44n, 102
Naval Personnel Research and Development Center (NPRDC), 71, 123, 144, 149, 150
Naval Reserve Officer Training Corps (NROTC), 65, 66, 67, 68, 69, 70, 71, 157, 203

Naval Technical Training,
  chief of, 122n
Naval Training,
  chief of, 23
NAVEXAMCEN. *See* Naval Examining Center.
Navy,
  affirmative action in, 16-30, 32-35
  equal opportunity in, 8-30, 32-35
  equal opportunity policy, 1-2, 6, 14-15
  Recruiting Command, 37, 39, 41, 42n, 45, 49n, 50, 51n, 54, 67, 75, 103, 104n, 105n, 163n, 166n
    assistant chief of personnel planning and programming, 49, 51n, 57n
  recruiting organization, 36-40, 186
Navy Affirmative Action Plan (NAAP) of 1976, 33-34
Navy Enlisted Classification (NEC), 44n
Navy Enlisted Occupational System (NEOCS), 154
Navy Race Relations Program, 22-30, 32-33
  Phase I, 22-27, 32, 33, 199
  Phase II, 27-30, 32, 33
Navy Race Relations School, 22-23
NEC. *See* Navy Enlisted Classification
Nelson, Dennis D., 8n, 9n, 10n, 11n
NEOCS. *See* Navy Enlisted Occupational System
neutral rating, 75, 82n, 151, 169. *See also* Career Reenlistment objectives
NFO. *See* naval flight officers
NJP. *See* nonjudicial punishment
NO. *See* Numerical Operations
nonjudicial punishment (NJP), 19, 20, 136, 148-49, 150, 151, 178
  and command's organizational effectiveness, 198-99
Norfolk (Virginia), 16n, 107n, 155n
North Island, 21
Northrup, H. R., 6n, 118n, 129n
NPRDC. *See* Naval Personnel and Development Center
NROTC. *See* Naval Reserve Officer Training Corps
nuclear occupations, 104, 127n, 130-31, 132-33, 165, 179-80, 181, 182-83, 204, 206
Nugent, William A., 124n

Numerical Operations (NO), 111
occupational classification, 95-133, 187-88
  after the draft, 100-105
  during the draft, 99-100
  of enlisted, 99-105
  objectives of, 100, 101
  of officers, 127-33, 204
  role of mental aptitude testing in, 187-88
OCS. See Officer Candidate School
Office of Federal Contract Compliance Programs (OFCCP), 13, 117
  General Order No. 4, 13-14
officer advancement. See advancement system, procedure for officers
Officer Candidate School (OCS), 65, 67, 68, 69, 70, 183, 203, 206
officer career pattern. See advancement system, role of career pattern in
officer entry routes, 65-71
officer fitness reports. See advancement system, role of fitness reports in
Oklahoma City, 16n
Omaha (Nebraska), 41
on-the-job training, 105-10, 193
open contract assignment, 102-04
open rating, 75, 82, 85-86, 87, 151-54 passim, 169-70. See also Career Reenlistment Objectives
operational command, 102n
Orlando (Florida), 41
other minorities model. See advancement function, other minorities model

Pacific front, 12
Parker, Jim, 170n
Parker, Warrington S., Jr., 150n
Parris Island, 38
passed-but-not-advanced (PNA) points, 138, 142
Patrick Air Force Base, 22
Pensacola (Florida), 16n, 43
performance evaluation model. See advancement function, performance evaluation model
performance evaluations, 136, 137
  and face-to-face counseling, 140-41
Philadelphia, 16n, 39, 107, 155n
Philippines, 11
Planter, 10
Plessy v. *Ferguson*, 10
Populists, 10

preservice model. See advancement function, preservice model
preservice plus in-service model. See advancement function, preservice plus in-service model
President's Committee on Equality of Treatment and Opportunity in the Armed Services, 13
private sector,
  manpower practices of, 109-10
  recruiting organization of, 36-40
  upgrading practices, 156
programmed texts, 132
Project 100,000, 60-63
promotion. See advancement system
Puerto Rican, 74, 76n

Qualified Military Available Statistics, 45, 47
Quantico (Virginia), 155n

Race Relations Education Specialists (RRES), 22-23, 24, 27, 28
race relations programs,
  Department of Defense, 22
  Marine Corps, 30-32
  Navy, 22-30, 32-33
reading level,
  of average recruit, 123-24
Record, Jeffrey, 52n
recruiter,
  length of assignment, 44-45
  responsibilities, 43
  selection, 40-41
  training, 41-43
"recruiter rating," 40
recruiting, 36-71
  educational criteria, 51-53, 56-57, 58, 63-65
  facilities for, 45-46
  malpractices, 59-60
  of minorities. See minorities, recruiting of
  moral criteria, 49-50
  personnel, 47-48
  physical criteria, 49-50
  role of mental aptitude testing in, 187
recruiting mix, 51-53
Recruiting Officer Management Orientation School (ROMO), 43
recruiting/training relationship, 48-49
Reeg, Frederick, J., 59
reenlistment,
  eligibility requirements, 140, 167-71, 200-201

Index

incentives, 174-80, 206
  of enlisted, 167-80
  of officers, 180-83, 206
  officer incentives, 181
  role of career pattern in, 180
  role of fitness reports in, 180
  role of military justice in, 167
  role of occupational specialty in, 180
  role of performance evaluations in, 197
remedial education, 80, 123-27, 188-92. See also formal school training
  cost-effectiveness of, 126-27, 190
  current programs, 124-29
Required Workshops Track, 28, 29
retention. See reenlistment
Revolutionary War, 8
Richmond (Virginia), 41
Rickover, H. G., 132
Rimland, Bernard, 70n
Robertson, David W., 142n, 143n, 144n, 145n
ROMO. See Recruiting Officer Management Orientation School
Roseen, Darien, 99n, 100n, 101
Royle, Marjorie H., 143n, 145n
RRES. See Race Relations Education Specialists

"salt and pepper" teams, 24
San Diego, 16n, 19, 20, 30, 38, 41, 107n, 125, 155n
San Francisco, 16n, 41
Santiago, 10
SAT (Scholastic Achievement Test), 47
Saylor, John C., 119
Scarborough, Edward, 58, 59, 65n, 68, 69, 160n, 183n
Schneider, Stephen A., 64n, 133n
SCORE. See Selective Conversion and Reenlistment
Sea Cadets, 15
secretary of the navy, 23n, 24n, 37, 161
selection board, 161-62
Selective Conversion and Reenlistment (SCORE), 175-77, 178-79
Selective Reenlistment Bonus (SRB), 174-75
Selective Service Act of 1948, 49
Selective Service Law of 1940, 11
Selective Training and Reenlistment (STAR), 176, 177-79

self-paced learning, 192
Senate Armed Services Committee, 52, 60
Shop Information (SI), 111
Shop Practices Test (SHOP), 73-93 passim, 111-12, 212, 213, 214, 223, 227, 231, 232
Siegel, Arthur I., 53n, 80n, 105n, 121n, 146n
similar-item-difficulty test, 145, 196
Smalls, Robert, 10
Spanish-American War, 10
SRB. See Selective Reenlistment Bonus
staff, 37, 39-40
Standlee, Lloyd S., 119
STAR. See Selective Training and Reenlistment
Stillman II, Richard J., 9n, 10n, 11n, 13n
Stoddert, Benjamin, 9
"striker board," 107
striker procedures, 104-05, 159
Strong Vocational Interest Bank, 70
Subic Bay, 18, 19
submarine pay, 181, 206
Sullivan, John A., 118n
Super, D. E., 107n, 108n
Systems Development Corporation, 27

Taylor, Elain N., 62n
Technical Skills Enlistment Bonus, 103-04
T-groups, 23
Thomas, Edmund D., 149n, 150n
Thomas, Patricia J., 70n, 71n, 149n, 150n
Title VII. See Civil Rights Act of 1964, as amended 1972
Travis Air Force Base, 22
Truman, Harry S., 12

UCMJ. See Uniform Code of Military Justice
Understanding Personal Worth and Racial Dignity (UPWARD) Seminars, 23-25, 26
undesignated strikers, 102, 105-10, 193
  improving placement opportunities for, 108-09
  job selection process for, 106-08
Uniform Code of Military Justice (UCMJ), 16, 19, 148-49
United States Armed Forces Institute (USAFI), 127, 156-57

United States Employment Service, 46
upgrading. *See also* minorities, upgrading of
  defined, 6
"up-or-out," 167-69, 172, 201
UPWARD. See Understanding Personal Worth and Racial Dignity Seminars
Urban League of Philadelphia, 46-47
USAFI. *See* United States Armed Forces Institute
U.S.S. *Constellation*, 19-21
U.S.S. *Kitty Hawk*, 18-19

validation,
  of advancement examinations, 147, 194
  of aptitude tests, 117-18
Vietnam, 20, 90
Vineberg, Robert, 62n

War Department, 11

Ward, Samuel W., 149n
warfare specialty. *See* advancement system, role of warfare specialty in
War Manpower Commission, 11
Warner, John W., 21
War of 1812, 9
warrant officer rank, 27n
"whole-man" concept. *See* advancement system, "whole-man" concept
WK. *See* work knowledge
Wojdylak, Marcella, 143n
Wool, Harold, 61n
Work Knowledge (WK), 111, 112
World War I, 11
World War II, 11-12

Xerox Training Center,
  Marine recruiter training, 43

YMCA, 47

Zumwalt, Elmo R., 21-22

Racial Policies of American Industry Series

1. *The Negro in the Automobile Industry,* by Herbert R. Northrup. 1968
2. *The Negro in the Aerospace Industry,* by Herbert R. Northrup. 1968
3. *The Negro in the Steel Industry,* by Richard L. Rowan. 1968
4. *The Negro in the Hotel Industry,* by Edward C. Koziara and Karen S. Koziara. 1968
5. *The Negro in the Petroleum Industry,* by Carl B. King and Howard W. Risher, Jr. 1969
6. *The Negro in the Rubber Tire Industry,* by Herbert R. Northrup and Alan B. Batchelder. 1969
7. *The Negro in the Chemical Industry,* by William Howard Quay, Jr. 1969
8. *The Negro in the Paper Industry,* by Herbert R. Northrup. 1969
9. *The Negro in the Banking Industry,* by Armand J. Thieblot, Jr. 1970
10. *The Negro in the Public Utility Industries,* by Bernard E. Anderson. 1970
11. *The Negro in the Insurance Industry,* by Linda P. Fletcher. 1970
12. *The Negro in the Meat Industry,* by Walter A. Fogel. 1970
13. *The Negro in the Tobacco Industry,* by Herbert R. Northrup. 1970
14. *The Negro in the Bituminous Coal Mining Industry,* by Darold T. Barnum. 1970
15. *The Negro in the Trucking Industry,* by Richard D. Leone. 1970
16. *The Negro in the Railroad Industry,* by Howard W. Risher, Jr. 1971
17. *The Negro in the Shipbuilding Industry,* by Lester Rubin. 1970
18. *The Negro in the Urban Transit Industry,* by Philip W. Jeffress. 1970
19. *The Negro in the Lumber Industry,* by John C. Howard. 1970
20. *The Negro in the Textile Industry,* by Richard L. Rowan. 1970
21. *The Negro in the Drug Manufacturing Industry,* by F. Marion Fletcher. 1970
22. *The Negro in the Department Store Industry,* by Charles R. Perry. 1971
23. *The Negro in the Air Transport Industry,* by Herbert R. Northrup et al. 1971
24. *The Negro in the Drugstore Industry,* by F. Marion Fletcher. 1971
25. *The Negro in the Supermarket Industry,* by Gordon F. Bloom and F. Marion Fletcher. 1972
26. *The Negro in the Farm Equipment and Construction Machinery Industry,* by Robert Ozanne. 1972
27. *The Negro in the Electrical Manufacturing Industry,* by Theodore V. Purcell and Daniel P. Mulvey. 1971
28. *The Negro in the Furniture Industry,* by William E. Fulmer. 1973
29. *The Negro in the Longshore Industry,* by Lester Rubin and William S. Swift. 1974
30. *The Negro in the Offshore Maritime Industry,* by William S. Swift. 1974
31. *The Negro in the Apparel Industry,* by Elaine Gale Wrong. 1974

Order from: Kraus Reprint Co., Route 100, Millwood, New York 10546

STUDIES OF NEGRO EMPLOYMENT

Vol. I. *Negro Employment in Basic Industry: A Study of Racial Policies in Six Industries (Automobile, Aerospace, Steel, Rubber Tires, Petroleum, and Chemicals)*, by Herbert R. Northrup, Richard L. Rowan, et al. 1970. *

Vol. II. *Negro Employment in Finance: A Study of Racial Policies in Banking and Insurance*, by Armand J. Thieblot, Jr., and Linda Pickthorne Fletcher. 1970. *

Vol. III. *Negro Employment in Public Utilities: A Study of Racial Policies in the Electric Power, Gas, and Telephone Industries*, by Bernard E. Anderson. 1970. *

Vol. IV. *Negro Employment in Southern Industry: A Study of Racial Policies in the Paper, Lumber, Tobacco, Coal Mining, and Textile Industries*, by Herbert R. Northrup, Richard L. Rowan, et al. 1971. $13.50

Vol. V. *Negro Employment in Land and Air Transport: A Study of Racial Policies in the Railroad, Airline, Trucking, and Urban Transit Industries*, by Herbert R. Northrup, Howard W. Risher, Jr., Richard D. Leone, and Philip W. Jeffress. 1971. $13.50

Vol. VI. *Negro Employment in Retail Trade: A Study of Racial Policies in the Department Store, Drugstore, and Supermarket Industries*, by Gordon F. Bloom, F. Marion Fletcher, and Charles R. Perry. 1972. $12.00

Vol. VII. *Negro Employment in the Maritime Industries: A Study of Racial Policies in the Shipbuilding, Longshore, and Offshore Maritime Industries*, by Lester Rubin, William S. Swift, and Herbert R. Northrup. 1974. *

Vol. VIII. *Black and Other Minority Participation in the All-Volunteer Navy and Marine Corps*, by Herbert R. Northrup, Steven M. DiAntonio, John A. Brinker, and Dale F. Daniel. 1979. $18.50

OTHER COLLECTIVE BARGAINING STUDIES

Open Shop Construction, by Herbert R. Northrup and Howard G. Foster. Major Study No. 54. 1975. $15.00

Restrictive Labor Practices in the Supermarket Industry, by Herbert R. Northrup and Gordon R. Storholm, Major Study No. 44. 1967. $7.50

Order from the Industrial Research Unit
The Wharton School, University of Pennsylvania
Philadelphia, Pennsylvania 19104

* Order these books from University Microfilms, Inc., Attn: Books Editorial Department, 300 North Zeeb Road, Ann Arbor, Michigan 48106.

MANPOWER AND HUMAN RESOURCES STUDIES

3. *Manpower in Homebuilding: A Preliminary Analysis*, by Howard G. Foster. 1974. $6.95
4. *The Impact of Government Manpower Programs*, by Charles R. Perry, Bernard E. Anderson, Richard L. Rowan, and Herbert R. Northrup. 1975. $18.50
5. *Manpower and Merger: The Impact of Merger Upon Personnel Policies in the Carpet and Furniture Industries*, by Steven S. Plice. 1976. $7.95
6. *The Opportunities Industrialization Centers: A Decade of Community-Based Manpower Services*, by Bernard E. Anderson. 1976. $7.95
7. *The Availability of Minorities and Women for Professional and Managerial Positions, 1970-1985*, by Stephen A. Schneider. 1977. $25.00
8. *The Objective Selection of Supervisors*, by Herbert R. Northrup, Ronald M. Cowin, Lawrence G. Vanden Plas, et al. 1978. $25.00
9. *Manpower in the Retail Pharmacy Industry*, by Herbert R. Northrup, Douglas F. Garrison, and Karen M. Rose. 1979. $9.50

MULTINATIONAL INDUSTRIAL RELATIONS SERIES

1. *Case Studies of Multinational Bargaining and Prospects*, by Herbert R. Northrup, Richard L. Rowan, et al. (Reprint collection of thirteen published articles covering thirteen industries.) 1974-1978. $10.00
2. *The Reform of the Enterprise in France*. Official English Translation of the "Sudreau Report." 1975. $10.00
3. *German Codetermination Act of May 4, 1976, and Shop Constitution Law of January 15, 1972*. English Translation. 1976. $10.00
4. Latin American Studies
   (4a—Brazil). *The Political, Economic, and Labor Climate in Brazil*, by James L. Schlagheck. 1977. $12.00
   (4b—Mexico). *The Political, Economic, and Labor Climate in Mexico*, by James L. Schlagheck. 1977. $12.00
   (4c—Peru). *The Political, Economic, and Labor Climate in Peru*, by Nancy R. Johnson. 1978. $12.00
   (4d—Venezuela). *The Political, Economic, and Labor Climate in Venezuela*, by Cecilia M. Valente. 1979. $12.00
   (Future studies in this series will cover Colombia, Argentina, and Chile.)

INDUSTRY STUDIES

Prescription Drug Pricing in Independent and Chain Drugstores, by Jonathan P. Northrup. 1975. $5.95

Market Restraints in the Retail Drug Industry, by F. Marion Fletcher. Major Study No. 43. 1967. $10.00

The Carpet Industry: Present Status and Future Prospects, by Robert W. Kirk. Miscellaneous Report Series No. 17. 1970. $5.95

The Economics of Carpeting and Resilient Flooring: An Evaluation and Comparison, by George M. Parks. Major Study No. 41. 1966. $2.95

SPECIAL REPRINT EDITION

Productivity Accounting, by Hiram S. Davis. A 1978 reprint edition of an outstanding, prescient analysis of the problem of defining and measuring productivity. Reprinted with the cooperation of the American Productivity Center, Inc. Foreword by Professor John W. Kendrick, foremost authority today on productivity. 1955, 1978. $15.00

Order from the Industrial Research Unit
The Wharton School, University of Pennsylvania
Philadelphia, Pennsylvania 19104